DISORDERS OF
LIPID METABOLISM

DISORDERS OF LIPID METABOLISM

Guido V. Marinetti

University of Rochester Medical Center
Rochester, New York

PLENUM PRESS • NEW YORK AND LONDON

Library of Congress Cataloging-in-Publication Data

Marinetti, Guido V. (Guido Vincent), 1918-
    Disorders of lipid metabolism / Guido V. Marinetti.
        p.   cm.
    Includes bibliographical references.
    ISBN 0-306-43431-8
    1. Lipids--Metabolism--Disorders.   I. Title.
RC632.L5M37 1990
616.3'997--dc20   89-72205
                                                              CIP

© 1990 Plenum Press, New York
A Division of Plenum Publishing Corporation
233 Spring Street, New York, N.Y. 10013

Printed in the United States of America

I dedicate this book to
my wife Antoinette,
my children Timothy and Hope,
and my other family members and friends

PREFACE

For the past 30 years I have been teaching lipid biochemistry to medical students, graduate students, and undergraduate students. The major topics covered in my courses were fatty acids, prostaglandins, leukotrienes, phospholipids, glycolipids, triacylglycerols, cholesterol, bile acids, and plasma lipoproteins. Emphasis was placed on the regulation and disorders of lipid metabolism. The latter included hyperlipidemias, atherosclerosis, and alcohol-induced liver damage.

In this volume, I have chosen to focus on the disorders of lipid metabolism at a level appropriate both for medical students and for graduate and undergraduate students majoring in the biological sciences. The biochemistry, nutrition, genetics, and cell biology aspects of lipids and lipid metabolism will be covered as they relate to lipid disorders. I am not aware of any textbook that integrates the disorders of lipid metabolism in this manner.

Chapter 1 includes a brief discussion of the basic structures, properties, and metabolism of lipids. This chapter is not very detailed, since the material covered is available in basic textbooks on biochemistry. The major focus of this volume is the various lipid disorders, with emphasis on polyunsaturated fatty acids, the molecular biology and pathogenesis of the hyperlipidemias, dietary and drug therapy for the hyperlipidemias, and alcohol-induced liver damage. The material presented has been obtained from several textbooks on biochemistry and from a variety of recent articles in the scientific literature.

Although written primarily for medical students, this book should be appealing to some graduate and undergraduate students majoring in biochemistry, physiology, pharmacology, genetics, and biology. It may also be of interest to practicing physicians, to nutritionists, and to some of the lay public. Most of the topics are covered in a broad, general manner. The book may be considered an intermediate-level textbook on lipid disorders and can serve as an adjunct to basic textbooks of biochemistry.

CONTENTS

Chapter 8
Drug Therapy for Hyperlipidemias: Lipid-Lowering Drugs and Antithrombic and Fibrinolytic Drugs 169

Chapter 9
Lung Surfactant Deficiency: Respiratory Distress Syndrome of the Newborn 189

INTRODUCTION

Lipids are essential chemicals for all living cells where they function as important structural components of cell membranes, as a primary source of energy, and as precursors for the biologically potent steroid hormones, prostaglandins, leukotrienes, and lipoxins. Lipids represent a variety of water- insoluble hydrophobic molecules. The major lipids are the fatty acids, phospholipids, neutral glycerides, glycolipids, sterols, sterol esters, and bile acids. These compounds differ considerably in their chemical and physical properties. Some, like the acylglycerols (glycerides), sterols, sterol esters, and glycosylceramides, are relatively nonpolar neutral molecules with very limited solubility in water. Others, such as the phospholipids, gangliosides, sulfatides, bile acids, and free fatty acids, are polar ionic molecules with detergentlike properties. Most of the lipids in living cells are complexed with proteins and make up the plasma lipoproteins and cell membranes. Plasma lipoproteins are the vehicles for transporting lipids in the blood.

Lipids in cells are constantly being synthesized and degraded so that their steady-state levels are maintained within certain limits. When one or more enzymes involved in lipid metabolism are defective or missing, the various lipid disorders can develop. Some of these disorders are listed in the following sections.

COMMON LIPID DISORDERS

Hyperlipoproteinemias–Atherosclerosis

There are five major types of hyperlipoproteinemias. Elevated levels of low-density lipoproteins (LDL) are associated with elevated blood cholesterol and give a greater risk of atherosclerosis, coronary heart disease, and stroke.

Obesity

Obesity, the accumulation of excess fat in adipose tissue, is accompanied by moderate to severe overweight. Afflicted persons have a higher risk of heart disease and other metabolic disorders.

Fatty Liver and Liver Cirrhosis

Fatty liver and liver cirrhosis are caused by the chronic excess intake of alcohol. Fatty liver is the accumulation of fat (triacylglycerols) in the liver and can be reversed by abstinence from alcohol. Liver cirrhosis, which afflicts about 30% of alcoholics, causes irreversible damage to liver cells and is associated with ammonia toxicity, ascites, esophageal varices, and bleeding.

Cholelithiasis

Cholelithiasis is the accumulation of gallstones in the gallbladder or bile ducts. This condition can lead to obstruction of the bile duct and inflammation of the gallbladder. Cholesterol stones form as a consequence of supersaturation of cholesterol in bile.

Ketoacidosis

Ketoacidosis is the excess production of ketone bodies and lowering of the pH of the blood. Ketoacidosis occurs in uncontrolled diabetes, prolonged starvation, and alcoholism. The major ketone bodies are acetoacetic acid and β-hydroxybutyric acid. Excess accumulation of lactic acid, propionic acid, methylmalonic acid, and propionic acid also cause acidosis.

Lung Surfactant Deficiency: Respiratory Distress Syndrome of the Newborn

Lung surfactant deficiency is a deficiency of a phospholipid–protein complex in the lung that lowers the surface tension in alveoli and prevents collapse of the lung. This abnormality affects about 25,000 premature babies per year in the United States.

Lipid Maldigestion and Malabsorption

Lipid maldigestion results from a lack of pancreatic lipase, colipase, or bile acids or from excessive acid production or low bicarbonate production, which inhibits pancreatic lipase. A deficiency of pancreatic lipase or colipase may occur in pancreatitis or in obstruction of the pancreatic duct. Bile acid deficiency may be caused by obstruction of the bile ducts.

Lipid malabsorption is due to abnormal enterocytes, as occurs in sprue, or to a deficiency of enterocytes as a result of surgical removal of some of the small intestines. Lipid malabsorption also can lead to a deficiency of the fat-soluble vitamins A, D, E, and K and to a deficiency of the essential fatty acids linolenic and linoleic acid.

RARE DISORDERS

Sphingolipidoses

Sphingolipidoses are caused by deficiencies of specific enzymes that are involved in the degradation of sphingolipids, including sphingomyelin, cerebrosides, gangliosides, and sulfatides. These diseases include Tay-Sachs disease, Sandhoff's disease, Niemann-Pick disease, Gaucher's disease, Farber's disease, Fabry's disease, and metachromatic leukodystrophy.

Refsum's Disease

The condition known as Refsum's results from a deficiency of phytanic acid oxidase, which leads to accumulation of the methyl-branched phytanic acid (3,7,11,15-tetramethylhexadecanoic acid) in cell membranes and causes brain damage.

Carnitine Deficiency

A deficiency of carnitine leads to impaired entry of fatty acids into mitochondria, which decreases fatty acid oxidation and causes triacylglycerol accumulation in cells.

Carnitine–Fatty Acyl-CoA Transferase Deficiency

A deficiency of carnitine–fatty acyl-CoA transferase produces symptoms very similar to those of carnitine deficiency since this enzyme is responsible for the entry of fatty acids into mitochondria.

Essential Fatty Acid Deficiency

A lack of linoleic or linolenic acid in the diet causes water loss in the skin, skin lesions, leaky red blood cells, loss of hair, weight loss, kidney damage, sterility, and possibly death.

Propionic Acid Aciduria

A deficiency of propionyl-CoA carboxylase leads to an accumulation of propionic acid in the blood and urine and can lead to acidosis.

Methylmalonic Acid Aciduria

A deficiency of methylmalonyl-CoA mutase leads to an accumulation of methylmalonic acid in the blood and urine and can lead to acidosis.

Lecithin-Cholesterol Acyltransferase Deficiency

Abnormal high-density lipoproteins (HDL) in the blood, abnormal red blood cells, anemia, kidney failure, and cholesterol accumulation in tissues can result from a deficiency of lecithin-cholesterol acyltransferase (LCAT).

Apolipoprotein A Deficiency

A deficiency of the protein apolipoprotein A (apoA) causes Tangiers disease which is characterized by low levels of HDL in the blood, elevated levels of very-high-density lipoproteins (VHDL), cholesterol accumulation in tissues, and enlarged tonsils.

Apolipoprotein B Deficiency

A deficiency of apoB leads to a decreased ability of the intestines to form chylomicrons and a decreased ability of the liver to form VLDL.

Apolipoprotein C-II Deficiency

A deficiency of apoC-II leads to impaired functioning of lipoprotein lipase so that chylomicrons and VLDL accumulate in the blood, leading to elevated blood triacylglycerols.

Deficiency of Pancreatic Lipase or Colipase

A deficiency of pancreatic lipase or colipase leads to impaired digestion of dietary triacylglycerols, steatorrhea, and malabsorption of dietary fat.

Apolipoprotein E Defects

A defect of the protein apoE leads to the accumulation of abnormal β-VLDL in the blood, which leads to premature atherosclerosis and coronary artery disease.

Deficiency of the LDL Receptor

A deficiency of the LDL receptor leads to elevated levels of cholesterol in the blood which leads to premature atherosclerosis and coronary artery disease.

Wolman's Disease

Wolman's disease is caused by a deficiency of acid lipases a condition that leads to the accumulation of cholesterol esters and triacylglycerols in tissues.

Cholesterol Ester Storage Disease

In the rare familial disorder known as cholesterol ester storage disease, in which the liver is enlarged and contains high levels of cholesterol esters as a result of a deficiency of cholesterol esterase.

Cerebrotendinous Xanthomatosis

The lipid abnormality cerebrotendinous xanthomatosis leads to the accumulation of cholesterol and cholestanol in the nervous system and tendons. The primary enzymatic defect is a deficiency of a mitochondrial steroid, 26-hydroxylase involved in the pathway for the formation of bile acids from cholesterol. This leads to the accumulation of 7-α-hydroxy-4-cholesten-3-one, which can then be converted to cholestanol.

β-Sitosterolemia with Xanthomatosis

β-Sitosterolemia with xanthomatosis is a rare disorder associated with tendinous and tuberous xanthomas. Patients have high levels of plant sterols, especially β-sitosterol, in the plasma, red cells, adipose cells, and skin. The disease appears to be caused by an increased absorption of plant sterols in the intestines.

Pseudohomozygous Familial Hypercholesterolemia

Persons with the rare disorder called pseudohomozygous familial hypercholesterolemia have very high levels of serum cholesterol but normal levels of serum triacylglycerols. Cutaneous planar xanthomas are often present. Both parents of these patients have normal cholesterol levels. Restriction of dietary cholesterol and use of cholestyramine are very effective in treating this disorder which distinguishes it from type IIa hypercholesterolemia.

Zellweger's Syndrome

Zellweger's syndrome is a rare familial cerebro-hepato-renal disease affecting many tissues; it is lethal within the first year after birth. In this lipid disorder, the liver and kidneys lack peroxisomes which results in impaired synthesis of ether phospholipids and bile acids and the inability to oxidize fatty acids by an α-hydroxyacid oxidase. Other functions of peroxisomes that depend on catalase are also impaired. The characteristics of this disease indicate that peroxisomes are essential for normal mammalian development.

Vitamin Deficiencies

Deficiencies of biotin and pantothenic acid lead to defective lipid synthesis or oxidation. Biotin is a coenzyme for the enzyme acetyl-CoA carboxylase, the key rate-limiting enzyme in the synthesis of fatty acids. Pantothenic acid is necessary for the synthesis of CoA, which is vital for the metabolism of fatty acids, phospholipids, cholesterol, and bile acids.

In addition to describing more fully the disorders above, the chapters that follow briefly discuss obesity, the relationship of dietary lipids to cancer, and lipid peroxidation. It is hoped that the information presented will provide a broad perspective on the biochemical, nutritional, molecular biology, and clinical aspects of the disorders of lipid metabolism.

Chapter 1

DISORDERS OF LIPID DIGESTION AND ABSORPTION

1.1. INTRODUCTION TO STRUCTURES AND PROPERTIES OF LIPIDS

Disorders of lipid metabolism fall into two major categories according to their frequency of occurrence and whether they are genetically or environmentally determined. Some lipid disorders, such as atherosclerosis, develop over many years and are influenced by both genetic and environmental factors. Other disorders such as the sphingolipidoses, involve rather specific genetic mutations of one enzyme; these may begin in utero and are usually fatal early in life. Therefore, a wide spectrum of lipid disorders occurs in humans, affecting persons either early or late in life and being either fairly mild or highly fatal. To understand the disorders of lipid metabolism, it is appropriate first to consider the types and structures of the lipids that occur in living cells and to discuss briefly their general properties.

Lipids, proteins, carbohydrates, and nucleic acids are ubiquitous constituents of all living cells. Unlike proteins, carbohydrates, and nucleic acids, most of which are hydrophilic, lipids are generally hydrophobic. For this reason, lipids (also called fats) are extracted from cells by organic solvents such as methanol, acetone, chloroform, or ether or by mixtures of these solvents. A 1:1 or 2:1 mixture of chloroform–methanol is widely used in research on lipid. The major lipids in animal cells are phospholipids, sphingolipids, triacylglycerols (triglycerides), glycolipids, cholesterol, and bile acids. Other very important and physiologically active lipids that occur in very low concentrations in cells are prostaglandins, leukotrienes, and lipoxins.

Lipids such asphospholipids, sphingolipids, triacylglycerols, and glycolipids contain fatty acids that are covalently linked to the polar residue glycerol or sphingosine. Because of their hydrocarbon nature, most lipids are hydrophobic molecules and are not very soluble in water. This property of phospholipids and sphingolipids allows them to form bilayers in aqueous media and makes them uniquely suited to form membranes of living cells. However, lipids such as triacylglycerols and cholesterol are very insoluble in water and do not by themselves form bilayers. Therefore, they are packaged into lipoprotein complexes in the blood. The lipoproteins contain an oily lipid core of predominantly triacylglycerols and cholesterol esters and an outer membrane consisting of phospholipids, cholesterol, and proteins. The outer surface of the membrane contains polar hydrophilic groups that confer water solubility on the lipoproteins.

Cell membranes consists of a lipid bilayer containing phospholipid, glycolipid, and cholesterol. Proteins are embedded in the lipid bilayer. Membranes divide the cell into specialized compartments and act as semipermeable barriers to regulate the flux of nutrients, ions, and waste products. Membranes also contain special protein receptors that receive signals from the environment and transduce these signals into the cell.

Lipids, in particular triacylglycerols, make up about 40% of the total calories of the American diet. This amounts to 156 g of fat per day, of which 60% is animal fat and 40% is vegetable fat. However, with the recent interest in ω-3 polyunsaturated fatty acids and in attempts to increase the ratio of polyunsaturated to saturated fatty acids, Americans are consuming more vegetable and fish oils. The major dietary lipids include triacylglycerols (70–160 g/day), phospholipid (2–4 g/day), and cholesterol (0.3–1.0 g/day). The triacylglycerols and phospholipids contain a wide variety of different fatty acids.

The major fatty acids of animal and plant tissues are shown in Table 1-1; the structures of the fatty acids are shown in Fig. 1-1A. Fatty acids are amphophilic molecules containing a hydrophobic hydrocarbon side chain and a polar carboxyl group. The number of carbon atoms and number of double bonds in the side chain vary, which accounts for the large number of different fatty acids in nature. Table 1-1 lists the sources of the fatty acids along with the classification of the

Table 1-1. Some Major Fatty Acids of Animals and Plants

Fatty acid[a]	Number of C atoms	Number of CH=CH	Source
Palmitic	16	0	Animal and plants
Stearic	18	0	Animal and plants
Oleic, ω−9	18	1	Animal and plants
Linoleic, ω−6	18	2	Animal and plants
Linolenic, ω−3	18	3	Animal and plants
Arachidonic, ω−6	20	4	Animal and plants
Eicosapentenoic, ω−3	20	5	High in fish
Docosahexenoic, ω−3	22	6	High in fish

[a] ω−n refers to the number of C atoms from the most distal double bond to the methyl-terminal end of the fatty acid. CH=CH represents a double bond. The double bonds in the fatty acids listed have the *cis* configuration.

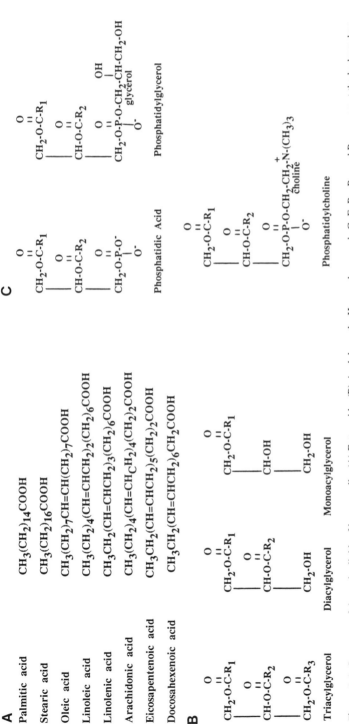

Figure 1-1. Structures of the major lipids of human cells. (A) Fatty acids. (B) Acylglycerols. Here and on panels C–F, R₁, R₂, and R₃, represent the hydrocarbon chains of the fatty acids. (C) Glycerolphospholipids. (D) Alkenyl glycerolphospholipids. (E) Sphingolipids. (F) Glycosylceramides. (G) Gangliosides. (From Devlin, 1982. Reproduced with permission.) (H) Cholesterol, cholesterol esters, and bile acids. (From Devlin, 1982. Reproduced with permission.) (I) Prostaglandins. (From Devlin, 1982. Reproduced with permission.) (J) Structures of lipoxins. (K) Leukotrienes.

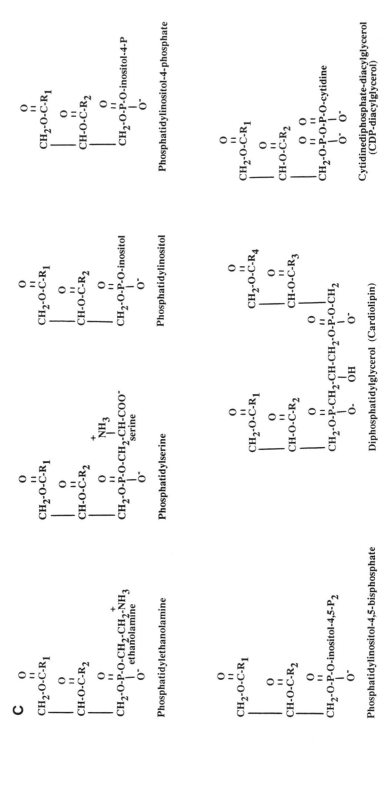

Figure 1-1. Continued

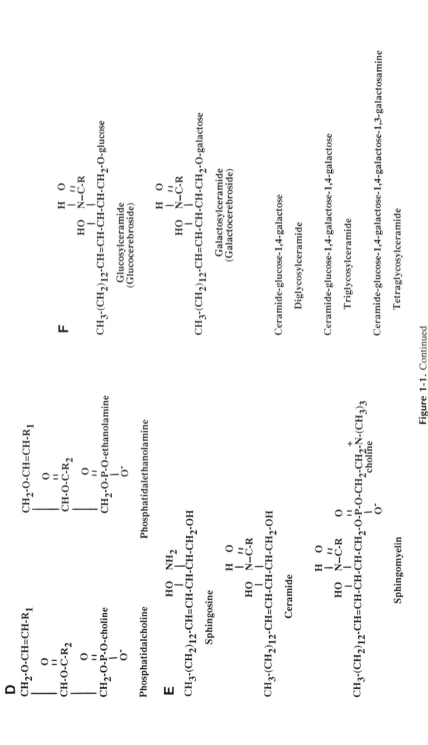

D

CH$_2$-O-CH=CH-R$_1$
|
CH-O-C-R$_2$
| ‖
| O
|
CH$_2$-O-P-O-choline
     ‖
     O
     |
     O$^-$

Phosphatidalcholine

CH$_2$-O-CH=CH-R$_1$
|
CH-O-C-R$_2$
| ‖
| O
|
CH$_2$-O-P-O-ethanolamine
     ‖
     O
     |
     O$^-$

Phosphatidalethanolamine

E

         HO   NH$_2$
         | |
CH$_3$-(CH$_2$)$_{12}$-CH=CH-CH-CH-CH$_2$-OH

Sphingosine

         H   O
         | ‖
         N-C-R
         |
HO
|
CH$_3$-(CH$_2$)$_{12}$-CH=CH-CH-CH-CH$_2$-OH

Ceramide

         H   O
         | ‖
         N-C-R
         |
HO   O
| ‖
CH$_3$-(CH$_2$)$_{12}$-CH=CH-CH-CH-CH$_2$-O-P-O-CH$_2$-CH$_2$-N-(CH$_3$)$_3$
                              | $^+$
                              O$^-$ choline

Sphingomyelin

F

         H   O
         | ‖
         N-C-R
         |
HO
|
CH$_3$-(CH$_2$)$_{12}$-CH=CH-CH-CH-CH$_2$-O-glucose

Glucosylceramide
(Glucocerebroside)

         H   O
         | ‖
         N-C-R
         |
HO
|
CH$_3$-(CH$_2$)$_{12}$-CH=CH-CH-CH-CH$_2$-O-galactose

Galactosylceramide
(Galactocerebroside)

Ceramide-glucose-1,4-galactose

Diglycosylceramide

Ceramide-glucose-1,4-galactose-1,4-galactose

Triglycosylceramide

Ceramide-glucose-1,4-galactose-1,4-galactose-1,3-galactosamine

Tetraglycosylceramide

Figure 1-1. Continued

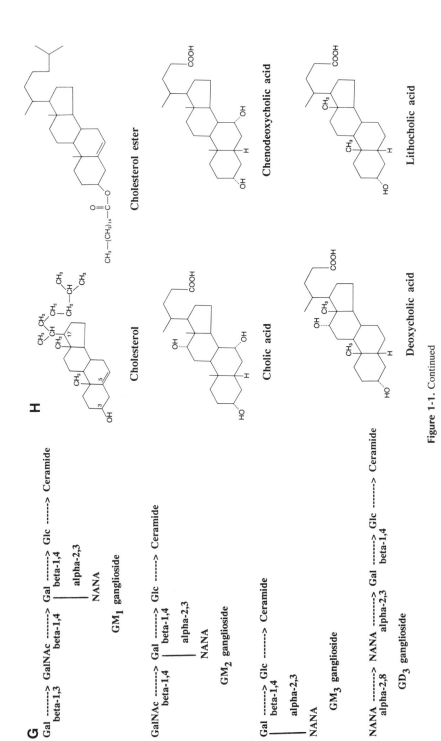

Figure 1-1. Continued

Figure 1-1. Continued

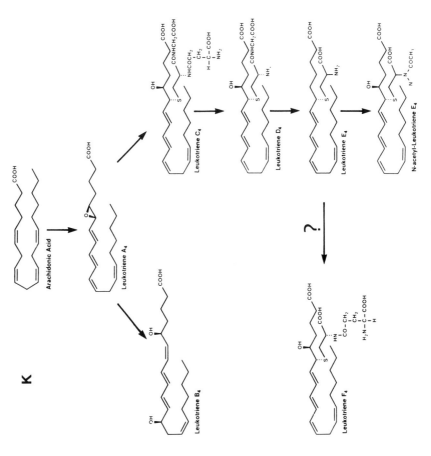

Figure 1-1. Continued

polyunsaturated fatty acids, using the ω designation. This designation is widely used today, especially among nutritionists, although some researchers use the letter "n" in place of ω.

Two major types of fatty acids, saturated and unsaturated, occur in nature. Saturated fatty acids have no double bonds and are solids at room temperature (if they have 12 or more carbon atoms), whereas unsaturated fatty acids have one or more double bonds (mainly of the *cis* configuration) and are liquids at room temperature. The fatty acids may have an even or odd number of carbon atoms and can be linear or branched. The most abundant fatty acids have an even number of 12–24 carbon atoms. The unsaturated fatty acids are further subdivided into two types: the monounsaturated fatty acids, which have one double bond, and the polyunsaturated fatty acids, which have two or more double bonds. The polyunsaturated fatty acids are further classified into three types, ω-3, ω-6, and ω-9, depending on the position of the double bond closest to the methyl-terminal (ω) end of the fatty acid chain. It is of particular interest that the ω-3 fatty acids occur in high amounts in fish whereas the ω-6 and ω-9 fatty acids occur in high amounts in vegetable and animal fats.

The interest in the polyunsaturated fatty acids arises from the fact that they are precursors of the biologically active prostaglandins, leukotrienes, and lipoxins. Prostaglandins play an important role in platelet aggregation and coronary heart disease. These topics are discussed in Chapters 6–8.

Nearly all of the fatty acids in animal cells are linked to more complex lipids in either an ester or amide bond or to CoA as a thiol ester bond. The major lipids that occur in living cells are the triacylglycerols, phospholipids, sphingolipids, glycolipids, cholesterol, and bile acids. The minor but very important lipids include prostaglandins, leukotrienes, and lipoxins. The lipid structures are shown in Figure 1-1. (Steroid hormones and fat-soluble vitamins which are also considered lipids will be discussed only very briefly in this book.)

Triacylglycerols (triglycerides), diacylglycerols (diglycerides), and monoacylglycerols (monoglycerides) are fatty acid esters of glycerol. They represent a family of lipids that differ in fatty acid composition. Triacylglycerols serve as a major storage form of fuel in animal cells and are stored in adipose tissue. The C-2 carbon atom of the acylglycerols has the sinister (*sn*) configuration.

Phospholipids (also called phosphatides) represent a variety of phosphorylated lipids that differ in fatty acid composition and in the polar organic molecules attached to the phosphate group. The "backbone" molecule to which fatty acids are linked is either glycerol or sphingosine. The other polar molecules attached to the phosphate group of the glycerolphospholipids are choline, ethanolamine, serine, inositol, inositolphosphates, and glycerol. The fatty acids that are esterified to the glycerolphospholipids are asymmetrically positioned. Saturated fatty acids occur primarily on the α (or C-1) position of glycerol, whereas unsaturated fatty acids occur primarily on the β (or C-2) position. The asymmetric carbon atom on the C-2 position of phospholipids has the *sn* configuration. According to the Rosanoff system, this configuration is designated by the letter L.

The most abundant phospholipids in animal cells are phosphatidylcholine and phosphatidylethanolamine. Phosphatidylserine and diphosphatidylglycerol (also called cardiolipin) occur in lesser amounts in most tissues. However, phosphatidylserine is relatively high in brain cells, and cardiolipin occurs primarily in the inner mitochondrial membrane. The phosphatidylinositols are minor constituents of cells, but they currently are receiving considerable attention, especially phosphatidylinositol-4-phosphate (PIP) and phosphatidylinositol-4,5-bisphosphate (PIP_2). PIP_2 has been shown to be the precursor of the two new "second messengers" diacylglycerol and inositol triphosphate. A specific membrane phospholipase C catalyzes the hydrolysis of PIP_2. This phospholipase is activated by certain hormones and growth factors that bind to specific membrane receptors.

Phospholipids of living cells are constantly being synthesized and degraded. The degradation of phospholipids is mediated by specific phospholipases that occur in different parts of the cell. The various types of phospholipases that act on the glycerolphospholipids are shown in Fig. 1-2.

Phospholipase A_1 hydrolyzes the C-1-linked fatty acid whereas phospholipase A_2 hydrolyzes the C-2-linked fatty acid. These lipases are esterases. On the other hand, phospholipases C and D are phosphodiesterases and cleave the phosphodiester bond. The phospholipase A_2, which occurs in cell membranes, is of particular interest since it releases arachidonic acid and other polyunsaturated fatty acids such as eicosapentenoic acid that are precursors for the synthesis of prostaglandins, leukotrienes, and lipoxins.

Other types of glycerolphospholipids that occur in relatively high amounts in tissues such as brain, heart, and testes are the plasmalogens (now called

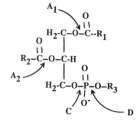

Figure 1-2. Different types of phospholipases acting on glycerolphospholipids. The phospholipases are indicated by A_1, A_2, C, and D. R_1, R_2, Hydrocarbon chains of the fatty acids; R_3, choline, serine, ethanolamine, inositol, and so forth.

phosphatidal lipids, alkenyl phospholipids, or alkanyl phospholipids). The alkenyl phospholipids contain a long-chain fatty aldehyde linked to the C-1 of glycerol as a vinyl ether bond. In the alkanyl phospholipids, the double bond of the vinyl ether is reduced to form a saturated ether linkage. The major phosphatidal lipids are phosphatidalcholine and phosphatidalethanolamine. The function of these lipids is not well understood. It is, however, known that platelet-activating factor, which contains an acetyl group on the C-2 position and a hexadecyl or octadecyl alcohol group on the C-1 position of glycerol, stimulates platelets to release biologically active amines such as epinephrine and serotonin.

Another major type of phospholipid is sphingomyelin. This lipid contains a fatty acid linked by an amide bond to the amino group of sphingosine as well as phosphorylcholine linked to a hydroxyl group of sphingosine. The fatty acids on sphingomyelin are longer chain and more saturated than the fatty acids on the glycerolphospholipids. Sphingomyelin is a major component of myelin, which surrounds certain brain and nerve cells.

The glycolipids are sphingolipids that contain one or more sugar units. The glycolipids include cerebrosides, sulfatides, and gangliosides and occur in high amounts in brain and nerve cells. The two types of cerebrosides are the glucocerebrosides, which contain glucose, and the galactocerebrosides, which contain galactose. Sulfatides contain a sulfate group on the sugar residue. Gangliosides also are of several types, depending on the number and type of sugar residues present. Mono-, di-, and trisialogangliosides contain one, two, and three sialic acids (neuraminic acid), respectively. They are abbreviated GM, GD, and GT, respectively, where M stands for monosialo, D stands for disialo, and T stands for trisialo; sialo refers to sialic acid, now more commonly called neuraminic acid. The numeral in the subscript of each ganglioside designation is calculated from $5-n$, where n is the number of neutral sugar residues in the ganglioside.

The major function of phospholipids and glycolipids is formation of asymmetric bilayer structures in cell membranes. The bilayer, which also contains cholesterol, acts as a barrier to the permeation of small polar molecules and allows for a fluid lipid environment in which are embedded specific membrane proteins. The proteins, which can be enzymes, transport proteins, hormone receptors, or growth factor receptors, confer specific properties on the membrane.

The sterol that occurs in animal cells is cholesterol. It has a hydroxylated cyclopentanoperhydrophenanthrene ring system with an isooctane side chain. Cholesterol is a white, waxy, water-insoluble lipid that is transported in the blood and in cells by complexing with lipoproteins. The major degradation products of cholesterol are the bile acids that are produced in the liver and stored in the gallbladder.

1.2. DIGESTION AND ABSORPTION OF LIPIDS

1.2.1. Lingual and Gastric Digestion

The digestion of dietary triacylglycerols begins in the mouth by the action of lingual lipase, which is an acid lipase with a pH optimum of about 4. The stomach contains an acid lipase that also hydrolyzes triacylglycerols. These acid lipases hydrolyze triacylglycerols to diacylglycerols and fatty acid. The lingual and gastric lipases are more important in infants who consume large quantities of milk in which the triacylglycerols are highly emulsified. Furthermore, the milk increases the very low pH of the stomach to a more favorable range (pH 4.0) for the lipase. Lingual and gastric lipases are believed to account for about 10–20% hydrolysis of triacylglycerols in adults and about 40–50% in infants.

Fat absorption is very efficient. Most fat is digested in the proximal region of the small intestine but with heavy intake of fat the absorption may extend to the distal region. The more saturated fats are digested and absorbed less well than the unsaturated fats.

Gastric lipase is produced by gastric glands in the stomach. This enzyme hydrolyzes triacylglycerols to diacylglycerols and fatty acids. Its release is stimulated by the secretagogue cholecystokinin (CCK), as depicted in Fig. 1-3.

CCK contains 33 amino acids, of which the carboxyl-terminal pentapeptide is identical to gastrin. It has a sulfated tyrosine-27, which is essential for full biological activity. CCK is identical to pancreozymin and is sometimes called CCK-pancreozymin. The synthetic octapeptide of CCK is five times more potent than the complete CCK. CCK is released from the upper small intestine. The most important stimuli for its release are the products of lipid digestion, in

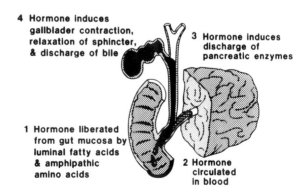

4 Hormone induces
gallbladder contraction,
relaxation of sphincter,
& discharge of bile

3 Hormone induces
discharge of
pancreatic enzymes

1 Hormone liberated
from gut mucosa by
luminal fatty acids
& amphipathic
amino acids

2 Hormone
circulated
in blood

Figure 1-3. Release and action of CCK. (From Patton and Hofmann, 1986. Reproduced with permission.)

particular fatty acids and monoacylglycerols. Amino acids and peptides derived from protein digestion also stimulate the release of CCK.

The fatty acids help emulsify the fats in the stomach. When they are ejected into the duodenum, they enhance the binding of pancreatic lipase to fat droplets, a process that allows pancreatic lipase to hydrolyze the triacylglycerols in the fat droplets. The fatty acids also induce the release of CCK from intestinal I cells. CCK then induces the gallbladder to contract and relaxes the sphincter of Oddi, allowing bile to flow into the small intestine. It also stimulates the secretion of pancreatic enzymes, potentiates the action of secretin in stimulating pancreatic bicarbonate secretion, inhibits gastric emptying, and stimulates bicarbonate secretion by the gallbladder. These events allow the intestinal phase of digestion to proceed.

Secretin is a 27-amino acid polypeptide that is released primarily from the duodenum. The major stimulus for its secretion is acidification of the small intestine. Secretin stimulates the secretion of NaCl and $NaHCO_3$ by the pancreas. Acetylcholine and CCK stimulate secretion of NaCl and digestive enzymes by the pancreas. Acetylcholine also stimulates the secretion of gastric and salivary juice.

1.2.2. Intestinal Digestion

The major site of digestion of dietary triacylglycerols is the small intestine. As mentioned above, hormones such as CCK and secretin influence lipid digestion by acting on the gallbladder or the pancreas. Stimulation of the vagus nerve also stimulates the secretion of pancreatic enzymes and contraction of the gallbladder. Local stimulation of I cells in the intestine by fatty acids and amino acids leads to the release of CCK, which enhances secretion of pancreatic enzymes and secretion of bile. In the duodenum the lipids are emulsified by bile acids in bile and the acidic stomach contents are neutralized by pancreatic juice, which is rich in bicarbonate.

The major phospholipid in bile is phosphatidylcholine, also called by the trivial name lecithin. The major bile acids in bile are cholic acid and chenodeoxycholic acid and their conjugated derivatives. Bile is released into the intestines about 5–10 times per day, during which it transports about 15–50 g of bile acids via the enterohepatic system.

Pancreatic juice contains two lipases: pancreatic lipase, which acts on triacylglycerols, and phospholipase A_2, which acts on certain phospholipids, particularly phosphatidylcholine. Pancreatic lipase has a pH optimum of 7 and requires colipase for full activity. Colipase is a small protein (mol. wt. 10,000) that is required for the binding of pancreatic lipase (mol. wt. 52,000) to emulsified triacylglycerol particles. Pancreatic lipase hydrolyzes the fatty acids linked to the 1 and 3 positions of the triacylglycerols and yields 2-monoacylglycerols lipase

Triacylglycerols + H$_2$O ---------> monoacylglycerols + fatty acids

colipase

bile acids

Figure 1-4. Pancreatic lipase hydrolysis of triacylglycerols in small intestine.

and free fatty acids (Fig. 1-4). This lipase differs from gastric lipase, which hydrolyzes triacylglycerols to diacylglycerols and free fatty acids. The fatty acids are neutralized with Na$^+$ and Ca^{2+} in the small intestines to form soaps, which together with bile salts help disperse and solubilize the undigested triacylglycerols (Fig. 1-5). The acidic bile salts and fatty acids occur on the outer surface of the micelle, where they confer negative charge repulsion on these lipid particles.

The bile salts represent a complex mixture of several conjugated and non-conjugated bile acids, including cholic acid, chenodeoxycholic acid, lithocholic

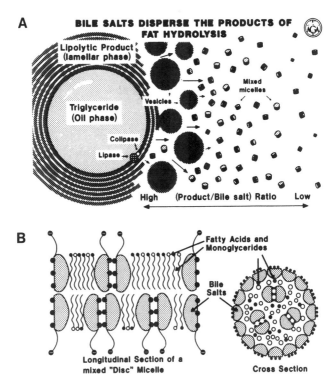

Figure 1-5. (A,B) Dispersion and solubilization of products of lipid digestion by bile acids. (From Patton and Hofmann, 1986. Reproduced with permission.)

acid, deoxycholic acid, and the taurine and glycine conjugates of these bile acids. The liver makes the primary bile acids (cholic and chenodeoxycholic acids) from cholesterol and conjugates these bile acids with taurine and glycine. The conjugated bile acids are more polar and stronger anions and thus are better emulsifying agents. The bacteria in the small intestines deconjugate and dehydroxylate the primary bile acids to form deoxycholic acid and lithocholic acid, which are considered secondary bile acids.

The estimated pool size (liver, intestine, and portal vein) of bile acids in the human is 3–5 g. Since the gallbladder contracts 5–10 times per day, the hepatic secretion rate is about 15–50 g/day. Approximately 0.3–0.6 g/day is lost in the feces. The absorption of bile acids occurs by passive diffusion in the jejunum and by active transport in the ileum.

A schematic diagram of lipid digestion and absorption and of the entero-hepatic circulation is shown in Fig. 1-6. The major digestion occurs in the small intestine by the action of pancreatic lipases. Since the most abundant dietary lipid is triacylglycerol, the major lipid digestion products are monoacylglycerols and free fatty acids. Other lipases hydrolyze dietary phospholipids, glycolipids, and cholesterol esters.

The intolerance for dietary lipid in patients with biliary obstruction suggests that lipid digestion is retarded but not completely blocked. Apparently, the small amounts of soaps and monoacylglycerols formed by partial hydrolysis of some triacylglycerols in the mouth, stomach, and small intestines along with lysolecithin produced by the action of pancreatic phospholipase A_2 are sufficiently active as detergents to allow pancreatic lipase to function to some degree in hydrolyzing dietary triacylglycerols.

Pancreatic phospholipase A_2 is a small protein (mol. wt. 14,000) that hydrolyzes glycerol-containing phospholipids in the diet, converting them to lysophospholipids and free fatty acid. The action of this lipase on phosphatidylcholine is shown in Fig. 1-7. The fatty acid on the 2 position of the glycerol backbone is released. Since the major dietary phospholipid is phosphatidylcholine (lecithin), the major product formed is lysophosphatidylcholine (lysolecithin). As mentioned above, lysolecithin is a detergent that helps emulsify the dietary triacylglycerols. Other hydrolytic enzymes are believed to occur in the small intestines and pancreas, and these enzymes break down phospholipids to free glycerol, glycerolphosphate, choline, inositol, serine, and ethanolamine. Choline and inositol have been called lipotropic agents since in animals they help mobilize triacylglycerols from the liver and aid in preventing fatty liver in animals treated with liver poisons such as carbon tetrachloride. However, dietary choline or inositol supplements have little or no lipotropic action in humans. Because of this lipotropic function in animals, choline has attracted the attention of food faddists who have nicknamed lecithin "big L" and made many unfounded claims for its curative properties.

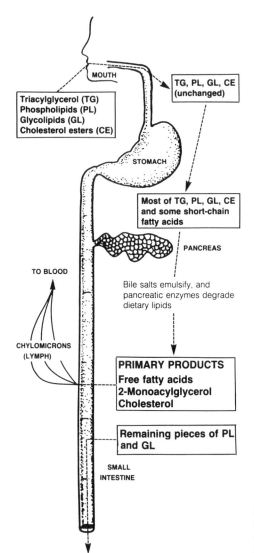

Figure 1-6. Overview of digestion and absorption of dietary lipids. (From Champe and Harvey, 1987. Reproduced with permission.)

$$\text{Phosphatidylcholine} + H_2O \xrightarrow{\text{phospholipase } A_2} \text{lysophosphatidylcholine} + \text{fatty acid}$$

Figure 1-7. Phospholipase A_2 hydrolysis of phosphatidylcholine in small intestine. (From Devlin, 1982. Reproduced with permission.)

Dietary cholesterol esters are hydrolyzed in the small intestine by pancreatic cholesterol esterase to yield free cholesterol and fatty acid. Bile salts are also required for emulsification of the water-insoluble cholesterol esters and for the subsequent emulsification of the free cholesterol and the hydrolysis products of lipid digestion (fatty acids and monoacylglycerols).

The products of lipid digestion (e.g., fatty acids, monoacylglycerols, cholesterol, lysophospholipids, glycerol, glycerolphosphate, choline, and inositol) are absorbed in the ileum and jejunum mainly by passive diffusion, very likely involving specific membrane transport proteins. Within the intestinal mucosal cells, the fatty acids and monoacylglycerols are converted back to triacylglycerols and phospholipids. These compouds, together with cholesterol, are packaged with specific proteins (mainly apoA and apoB-48) to form lipoprotein complexes called chylomicrons.

The chylomicrons are large lipoprotein particles with a diameter of 100–500 nm. These heterogeneous particles contain an outer monolayer membrane consisting of phospholipid, apoproteins, and cholesterol and a large inner core of triacylglycerols and cholesterol esters. The chylomicrons are called exogenous particles because they are made from dietary lipids. They are absorbed via the lymphatic system and enter the blood at the angle of the left jugular and subclavian veins. The chylomicrons give the blood a milky appearance, a condition called postprandial lipemia. The metabolism of chylomicrons is discussed in Chapter 5.

The triacylglycerols are digested by pancreatic lipase and 2-monoacylglycerol, and free fatty acids are the major products. Some glycerol is formed as a result of the action of an isomerase that converts the 2-monoacylglycerol to the 1-monoacylglycerol, which is hydrolyzed by pancreatic lipase to free glycerol and fatty acid. Within the intestinal mucosal cells, the monoacylglycerols, free fatty acids, and glycerol are converted back to triacylglycerols that are packaged into chylomicrons.

Fatty acids of chain length less than 10 carbon atoms and water-soluble products such as glycerol and choline are absorbed via the portal system and consequently go directly to the liver. Milk triacylglycerols have a relatively high content of fatty acids of shorter chain length. In contrast, vegetable and animal triacylglycerols contain very little or no shorter chain fatty acids.

Long-chain fatty acids (more than 14 carbon atoms) are reesterified to triacylglycerols and end up in chylomicrons. The conversion of these fatty acids to triacylglycerols occurs in mucosal cells via the monoacylglycerol pathway. Some fatty acids are also converted to phospholipids via the phosphatidic acid (glycerolphosphate) pathway. These reactions are shown in Fig. 1-8.

The lipid content of human plasma is shown Table 1-2. The lipid values are influenced by diet, hormones, age, sex, exercise, and other factors. Lipid analy-

A

Fatty acid + CoA + ATP -------> acyl-CoA + ADP

2Acyl-CoA + monoacylglycerol -----> triacylglycerol + 2CoA

B

Glycerol + ATP -----------> glycerolphosphate + ADP

Glycerolphosphate + 2acyl-CoA ---> phosphatidic acid + 2CoA

Phosphatidic acid + H_2O ----> diacylglycerol + Pi

Diacylglycerol + acyl-CoA ------> triacylglycerol + CoA

Figure 1-8. Two pathways for the synthesis of triacylglycerols in the small intestine. The mono-glyceride pathway (A) predominates in the small intestine whereas the phosphatidic pathway (B) predominates in liver.

sis of plasma should be done after an overnight fast. This procedure is especially important for analysis of plasma triacylglycerol.

After a very high fat meal, the triacylglycerol level of plasma can rise dramatically, reaching levels as high as 1000 mg/dl or more, depending on the amount of fat consumed and whether the person has a normal or defective

Table 1-2. Lipid Content of Human Plasma[a]

	Content (mg/dl)	
Lipid type	Mean	Range
Total lipid	570	360–820
Triacylglycerols	142	80–180
Total phospholipid	215	123–390
Phosphatidylcholine	175	50–200
Phosphatidylethanolamine	90	50–130
Sphingomyelin	25	15–35
Total cholesterol	200	107–320
Free cholesterol	55	26–106
Free fatty acids	12	6–16

[a]From Murray and Granner, 1988. Reproduced with permission. The mean values for phosphatidylcholine, phosphatidylethanolamine, and sphingomyelin are estimated averages of the range of values calculated by G. Marinetti.

lipoprotein lipase, or has a deficiency of the lipase. Due to the action of lipoprotein lipase the chylomicrons are degraded to small particles called chylomicron remnants as a result of the breakdown of the core triacylglycerols to free fatty acids and glycerol. The free fatty acids become bound to albumin and are carried to the liver and other tissues of the body. Chylomicrons have a half-life in plasma of about 15 min; because of their high concentration, however, it takes 6–8 hours for them to be cleared and for the triacylglycerol level in the blood to return to normal.

A cDNA for human lipoprotein lipase that codes for a protein of 448 amino acids has been cloned and sequenced. The sequence indicates that lipoprotein lipase, hepatic lipase, and pancreatic lipase are members of one gene family. The gene for human hormone-sensitive lipase has been mapped to chromosome 19 cent-q13.3 (Holm et al., 1988). The lipases which occur in lower animals and humans are listed in Table 1-3.

Most of the glycerol released by lipoprotein lipase is utilized, mainly by the liver, where it is phosphorylated by glycerol kinase and ATP to form glycerolphosphate. The latter is used to make triacylglycerols and phospholipids (Fig. 1-9), or can be converted to glucose via gluconeogenesis.

The fatty acids formed during the hydrolysis of triacylglycerols by lipoprotein lipase are used by most cells of the body (especially skeletal muscle, heart, kidney, and liver cells) for energy production. The excess fatty acids are re-esterified to triacylglycerols and stored as fat droplets in adipose tissue or are esterified to phospholipids. Insulin stimulates this process of fat storage.

During starvation or stress an adipose tissue lipase (which is different from lipoprotein lipase) is activated by several hormones, notably epinephrine and the

Table 1-3. Lipases of Human and Animal Tissues

Enzyme	Origin	Function
Gastric lipase	Stomach	Degrades triacylglycerols
Pancreatic lipase	Pancreas	Degrades triacylglycerols
Lipoprotein lipase	Extrahepatic tissues	Degrades triacylglycerols in chylomicrons and VLDL
Hormone-sensitive lipase	Adipocytes	Degrades triacylglycerols
Acid lipase	Many cells	Removes fatty acids from lipids taken up by lysosomes
Hepatic lipase	Liver	Degrades triacylglycerols in lipoprotein remnants

Glycerol + ATP ----------> glycerolphosphate + ADP

Glycerolphosphate + 2Acyl-CoA ----> phosphatidic Acid + 2CoA

Phosphatidic acid + H_2O ----------> diacylglycerol + Pi

Diacylglycerol + acyl-CoA ---------> triacylglycerol + CoA

Diacylglycerol + CDPC ------------> phosphatidylcholine + CMP

Diacylglycerol + CDPE -------> phosphatidylethanolamine + CMP

Figure 1-9. Synthesis of triacylglycerols and phospholipids in the small intestine. CDPC, Cytidinediphosphatecholine; CDPE, cytidinediphosphateethanolamine.

adrenocorticotropic hormone (ACTH). This lipase hydrolyzes the stored triacylglycerols and produces fatty acids, which enter the blood and are used by the other tissues of the body for energy. The action of epinephrine in releasing fatty acids from adipose tissue is called lipolysis. This action is mediated by membrane receptors, which bind the lipolytic hormones and, via specific GTP-binding proteins activate the enzyme adenyl cyclase, which converts ATP to cAMP. cAMP in turn activates a protein kinase that phosphorylates and activates the adipose tissue lipase. The lipase hydrolyzes the stored triacylglycerols and releases the fatty acids into the blood. The fatty acids bind to albumin in the blood and are delivered to cells, which require these fatty acids for energy. Recent studies have shown that cells contain specific membrane proteins that bind the fatty acids and enable their uptake into cells.

Insulin, an antilipolytic hormone, inhibits adenyl cyclase and activates a phosphodiesterase that degrades cAMP to 5'-AMP. This lowers the cAMP level in the cell and prevents the activation of protein kinase. Insulin also stimulates a phosphatase that dephosphorylates and "turns off" the adipose tissue lipase. On the other hand, insulin stimulates fatty acid synthesis in adipose tissue and liver and also stimulates the entry of glucose into adipocytes, which provides glycerolphosphate for the reesterification of the fatty acids to form triacylglycerols. Insulin is thus a lipogenic hormone.

Unlike triacylglycerols and phospholipids, which yield oxidizable degradation products (fatty acids and glycerol), cholesterol does not serve as a fuel for the body. Cholesterol functions as an important component of cell membranes, as a precursor for adrenal steroids, as a precursor for vitamin D in the skin, and as a substrate for the synthesis of bile acids in the liver.

1.3. CHARACTERISTICS OF DISORDERS OF LIPID DIGESTION AND ABSORPTION

Disorders of lipid digestion and absorption involve either defects in intraluminal processing of dietary lipids or impaired efficiency of absorption of the digested lipids.

The intraluminal defects include impaired hydrolysis, impaired micellar dispersion, and impaired pH regulation. Impaired hydrolysis may be due to a deficiency of pancreatic lipase or colipase. A deficiency of bile acids causes impaired micellar dispersion (emulsification) of the dietary triacylglycerols. Impaired pH regulation results from excess production of H^+ or a deficiency of bicarbonate. The resulting low pH inhibits pancreatic lipase.

Mucosal defects causing fat malabsorption may result from a decreased number of enterocytes (such as can occur in intestinal resection), from the presence of defective enterocytes (as occurs in sprue), or from decreased exposure of digestion products to the enterocytes (as with an intestinal bypass or fistula).

In diseases of the intestinal mucosa, lipid malabsorption can be severe without any impairment in lipid digestion. If lipid maldigestion is mild, there may be little or no lipid malabsorption.

Lipid maldigestion and malabsorption also can be caused by other factors including impaired or delayed release of CCK, drugs such as cholestyramine, which bind bile acids, and obstruction of the bile duct. Some causes of impaired hydrolysis of dietary triacylglycerols are listed in Table 1-4.

When bile is deficient as a result of severe liver dysfunction or biliary obstruction, lipid digestion and absorption are markedly impeded and lipid accumulates in the feces. This condition is called steatorrhea. The accumulated lipids consist of fatty acid salts and undigested triacylglycerols. Other lipid-soluble substances such as the fat-soluble vitamins A, D, E, and K also are poorly absorbed, a condition that can lead to vitamin deficiency. The consequences of pancreatic lipase and bile acid deficiency are presented in Table 1-5.

If the pancreas is damaged or the pancreatic duct is obstructed, pancreatic enzymes will not be made or will not be released, and a deficiency of pancreatic lipase and phospholipase A_2 occurs. The deficiency of pancreatic lipase or colipase gives rise to a fatty steatorrhea consisting mainly of undigested triacylglycerols. This condition causes a loss of important fuels and can lead to severe weight loss. When the lipase concentration is less than 10% of normal, lipolysis is severely impaired, and undigested triacylglycerols remain in the oil phase and are not absorbed. This condition causes a loss of energy to the body and also leads to malabsorption of the fat- soluble vitamins A, D, E, and K. If bacteria in the colon hydrolyze some of the triacylglycerols, a fatty acid diarrhea may occur.

**Table 1-4. Causes of Impaired
Triacylglycerol Hydrolysis**[a]

Deficiency	Examples
Low pancreatic lipase secretion	Chronic pancreatitis, cystic fibrosis, lipase deficiency
Normal lipase secretion with enzyme inactivation	
Excess gastric H^+	Gastrinoma
Decreased pancreatic bicarbonate	Chronic pancreatitis
Decreased colipase secretion	Colipase deficiency

[a]From Patton and Hofmann, 1986. Reproduced with permission.

Certain patients with islet cell tumors or antral hyperplasia have high levels of serum gastrin associated with gastric hypersecretion and acidic jejunal pH during digestion. This condition is called Zollinger-Ellison syndrome. The pancreatic lipase is inactivated at pH <3; in these patients, lipase activity is very low and triacylglycerol digestion markedly impaired.

Damaged mucosal cells (which can occur in celiac disease in children and sprue in adults) interfere with absorption of the products of lipid digestion, giving rise to steatorrhea and can lead to vitamin deficiency. In these cases, the major lipid in the feces consists of fatty acid salts.

**Table 1-5. Consequences of Pancreatic Lipase
and Bile Acid Deficiencies**[a]

Deficiency	Consequence
Pancreatic lipase	
Decreased fat absorption	Steatorrhea, decreased absorption
Decreased micellar lipid	of fat, and soluble vitamins
Bile Acid	
Inability to solubilize:	
Fat soluble vitamins	Vitamin deficiency
Fatty acids	Caloric deficiency
Monoacylglycerols	Increased fatty acids in the colon, steatorrhea, diarrhea

[a]From Patton and Hofmann, 1986. Reproduced with permission.

Decreased bile acid secretion may occur with parenchymal liver disease, obstructive hepatobiliary disease, and conditions interrupting the enterohepatic circulation of bile acids. The deficiency of bile acids leads to impaired micellar solubilization (emulsification) of the dietary lipids and their digested products and leads to lipid malabsorption.

Precipitation of bile acids occurs after administration of drugs such as neomycin and colestid (cholestyramine). These positively charged drugs bind the negatively charged bile acids and form insoluble complexes that prevent the reabsorption of the bile acids via the enterohepatic cycle. If given in excess, these drugs can interfere with lipid digestion and absorption.

Medium-chain triacylglycerols (which have fatty acids less than 12 carbon atoms long) are used as a source of dietary lipid for persons with bile acid deficiency, since the fatty acids and monoacylglycerols released by digestion of these lipids by pancreatic lipase do not require emulsification by bile acids and are absorbed directly into the vascular system rather than through the lymphatics.

In lymphangiectasia, the lacteals are dilated and filled with chylomicrons after a fatty meal. Some of the chylomicrons leak back into the intestinal lumen. In lymphoma or Whipple's disease, there is a cellular infiltrate around the lacteals and small lymphatics, causing obstruction of flow.

Diarrhea is also caused by fat malabsorption. Unabsorbed triacylglycerols alone do not cause diarrhea. However, microbial hydrolysis of triacylglycerols in the colon produces free fatty acids and hydroxylated fatty acids that have a direct stimulatory effect on the colonic mucosa, which leads to diarrhea.

Bile salt deficiency, lipase deficiency, and decreased or defective mucosa or lymphatics can lead to fat malabsorption and diarrhea. The normal or hydroxylated fatty acids induce net water secretion. Severe diarrhea leads to dehydration and acidosis, which can be life threatening, especially to infants and children. The acidosis results from a loss of Na^+ and K^+ from the small intestine, resulting in a lowering of the pH of the blood. When the pH of the blood falls below 7.0, coma and death can result.

The next chapter considers the disorders of fatty acid metabolism. Fatty acids are very important fuels for many cells of the body, particularly liver, skeletal muscle, heart muscle, and kidney. During prolonged starvation, when glycogen reserves are depleted, fatty acids and ketones are the major fuel for many cells of the body, including brain cells. Fatty acids are derived either from the diet or from de novo synthesis in the body. Since acetyl-CoA is the substrate for the de novo synthesis of fatty acids, any dietary food that can be converted to acetyl-CoA in the body can be converted to fatty acids. The major dietary foods, which when taken in excess are converted to fatty acids, are carbohydrates. The excess fatty acids are stored in adipose tissue as triacylglycerols, the primary storage form of fuel in the body.

REFERENCES

Booth, C. C., 1967, Sites of absorption in the small intestine, *Fed. Proc.*, 26:1583.

Borgstrom, B., Erlanson-Albertsson, C., and Wieloch, T., 1979, Pancreatic colipase: chemistry and physiology, *J. Lipid Res.*, 20:805.

Champe, P. C., and Harvey, R. A., (eds.), 1987, *Lippincott's Illustrated Reviews: Biochemistry*, J. B. Lippincott Co., Philadelphia.

Devlin, T.M. (ed.), 1982, *Textbook of Biochemistry with Clinical Correlations*, John Wiley & Sons, New York.

Feuerstein, G., and Hallenbech, J.M., 1987, Leukotrienes in health and disease, *Fed. Proc.*, 1:186.

Green, P.H.R., and Glickman, R.M., 1981, Intestinal lipoprotein metabolism, *J. Lipid Res.*, 22:1153.

Hofmann, A.F., 1970, Gastroenterology: physical events in lipid digestion and absorption, *Fed. Proc.*, 29:1317.

Holm, C., Kirchgessner, T.G., Svenson, K.L., Fredrickson, G., Nilsson, S., Miller, C.G., Shively, J.E., Heinzmann, C., Sparkes, R.S., Mohandas, T., Lusis, A.J., Belfrage, P., and Schotz, M.C., 1988, Hormone-sensitive lipase: sequence, expression, and chromosomal location to 19 cent-q 13.3, *Science*, 241:1503.

Isselbacher, K.J., 1967, Biochemical aspects of lipid malabsorption, *Fed. Proc.*, 26:1420.

McGilvery, R.W., and Goldstein, G.W., 1983, *Biochemistry, a Functional Approach*, W. B. Saunders Co., New York.

Moog, F., 1981, The lining of the small intestine, *Sci. Am.*, 245:154.

Murray, R.K., Granner, D.K., Mayes, P.A., and Rodwell, V.W., 1988, *Harper's Biochemistry*, 21st ed., Appleton & Lange, Norwalk, Connecticut.

Patton, J.S. and Hofmann, A.F.,1986, Lipid Digestion, Unit 19, in *The Undergraduate Teaching Project in Gastroenterology, Liver Disease*, p.14-15, 45-47, American Gastroenterological Association, Milner-Fenwick, Inc. (distributor), Timonium, Maryland.

Samuelsson, B., Dahlen, S., Lindgren, J.A., Rouzer, C.A., and Serhan, C.N., 1987, Leukotrienes and lipoxins: structures, biosynthesis, and biological effects, *Science*, 237:171.

Shields, H., Bates, M.L., Bass, N.M., Best, C.Y., Alpers, D. H., and Ockner, R.K., 1986, Light microscopic immunocytochemical localization of hepatic and intestinal types of fatty acid-binding proteins in rat small intestine, *J. Lipid Res.*, 27:549.

Smith, E.L., Hill, R.L., Lehman, I.R., Lefkowitz, R.J., Handler P., and White, A., 1983a, *Principles of Biochemistry: Mammalian Biochemistry*, 7th ed., McGraw-Hill Book Co., New York.

Smith, E.L., Hill, R.L., Lehman, I.R., Lefkowitz, R.J., Handler, P., and White, A., 1983b, *Principles of Biochemistry: General Aspects*, 7th ed., McGraw-Hill Book Co., New York.

Chapter 2

DISORDERS OF FATTY ACID METABOLISM

2.1. GENERAL ASPECTS OF FATTY ACID METABOLISM

Excess dietary fatty acids and excess dietary carbohydrate are stored in adipose tissue as triacylglycerols. Insulin is the major hormone that stimulates this process of fat storage. Fatty acids are a prime fuel for humans and lower animals; they can be stored as triacylglycerols in large amounts in fat cells.

The distribution of fuel storage in the body is shown in Table 2-1. It is evident from the table that fats, as triacylglycerols, represent a much higher storage of energy than do carbohydrates and proteins. During starvation, glycogen and triacylglycerol stores are used first to supply the energy needs of the body. Glycogen reserves last only about 2 days whereas fat reserves can last for 1–2 months, depending on the body fat content. Proteins, especially the vital ones, are oxidized last, since their depletion is a threat to life. However, some muscle proteins are oxidized during starvation. During starvation, muscle activity is greatly reduced and thus some muscle proteins can be sacrificed to meet the energy demands of the brain and heart. It is also noteworthy that fats yield more energy and more water per gram when they are oxidized in the body than do the other foods (Table 2-2).

The importance of fats as a prime storage form of energy is evident in nature. Migrating birds, hibernating animals, migrating seals, plant seeds, and eggs with a shell all contain high amounts of fats that serve the energy needs of the animal or cell. In migrating birds and other animals this energy is used for muscle work; for eggs and seeds it is needed for growth of the embryo.

When lower animals and humans require energy during stress or starvation, the stored fats in the body are mobilized via the action of certain hormones such as epinephrine and ACTH. These hormones bind to specific membrane receptors and activate the lipase in the fat cell. The lipase hydrolyzes the stored triacylglycerols and liberates fatty acids. The fatty acids enter the blood and are

31

Table 2-1. Estimated Fuel Reserves in a 70-kg Man[a]

Organ or tissue	Amount (kcal)		
	Proteins	Glucose	Triacylglycerols
Blood	0	60	45
Liver	400	400	450
Brain	0	8	0
Muscle	24,000	1200	450
Adipose tissue	40	80	135,000

[a]From Cahill, 1976. Reproduced with permission.

Table 2-2. Energy and Water Produced when Oxidized

Food	Amount (kcal/g)	Water/food (g/g)
Carbohydrate	4.2	0.4
Proteins	4.3	0.4
Fat	9.2	1.1
Ethanol	7.0	0.9

taken up by various tissues for energy production. The body tissues that use fatty acids as a major source of energy are heart muscle, skeletal muscle, kidney, and liver.

2.1.1. Oxidation of Fatty Acids

Fatty acids are carried in the blood bound to albumin. The fatty acids are transferred to specific transport proteins on the plasma membrane of cells and are transported into the cell. Some fatty acids may permeate the membrane by simple diffusion.

The metabolism of the fatty acids requires that they first be activated by conversion to CoA thiol esters (Fig. 2-1). This reaction is catalyzed by acyl-CoA

$$RCOOH + CoA\text{-}SH + ATP \longrightarrow RCOSCoA + AMP + PPi$$

Figure 2-1. Activation of fatty acids using ATP.

$$RCOOH + CoA\text{-}SH + GTP \text{------}> RCOSCoA + GDP + Pi$$

Figure 2-2. Activation of fatty acids using GTP.

synthases. These enzymes are named according to the length of the carbon chain of the fatty acid that reacts most rapidly; e.g., acetyl-CoA synthase acts fastest on C-2 and C-3 fatty acids; octanoyl-CoA synthase acts fastest on C-4 to C-12 fatty acids; and dodecanoyl-CoA synthase acts fastest on C-10 to C-18 fatty acids. These synthetic reactions occur in the endoplasmic reticulum and in the outer membrane of the mitochondria.

The matrix of the mitochondria also contains an enzyme that uses GTP rather than ATP for fatty acid activation (Fig. 2-2). This mitochondrial matrix enzyme probably is needed to activate free fatty acids that are produced in the mitochondria by lipases.

Another enzyme, which occurs mainly in extrahepatic tissues, is important for the activation of free acetoacetic acid, which is produced mainly in the liver during ketogenesis. This enzyme, succinyl-CoA-acetoacetate thiophorase (CoA transferase), converts acetoacetate to its CoA derivative (Fig. 2-3). This reaction is very important for the utilization of excess free acetoacetate produced in the liver during starvation and in uncontrolled diabetics. The enzyme acetoacetyl-CoA synthetase also activates free acetoacetic acid.

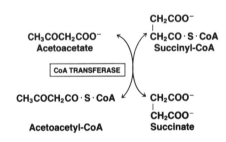

Figure 2-3. Conversion of free aceto-acetic acid to acetoacetyl-CoA. (From Murray et al., 1988. Reproduced with permission.)

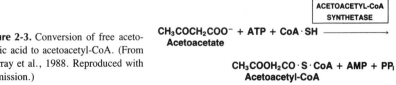

2.1.2. Role of Carnitine

Fatty acyl-CoA derivatives do not permeate the mitochondrial membrane. This creates a problem for the cell, since the enzymes that oxidize the fatty acids are located in the mitochondrial matrix. Carnitine (L-3-hydroxy-4-trimethyl-ammonium butyrate) serves as the carrier of the fatty acyl groups into the mitochondria (Fig. 2-4).

Acetyl-CoA carnitine transferase transfers acetic acid and other short-chain fatty acids through the inner mitochondrial membrane. Apparently, acetyl-CoA carnitine transferase does not play a major role in transporting acetyl-CoA groups in and out of the mitochondria, since this role is mediated primarily by the citrate shuttle. Carnitine palmityl-CoA transferase transfers long chain fatty acids through the inner mitochondrial membrane. This latter enzyme complex consists of at least two separate enzymes named CAT-I and CAT-II, which are asymmetrically localized on the outer and inner surfaces, respectively, of the inner mitochondrial membrane. A translocase shuttles carnitine and acyl carnitine across the membrane. A model of the carnitine shuttle is shown in Fig. 2-5.

2.1.3. Deficiency of CAT-I and Carnitine

A deficiency of muscle CAT-I has been observed in humans. This condition leads to muscle weakness, since the fatty acids in muscle cannot be oxidized sufficiently to meet the energy needs of the muscle cells. Moreover, since fatty acid oxidation is impaired, the fatty acids accumulate and are esterified to tri-acylglycerols. In muscle, the triacylglycerols accumulate as oily droplets that interfere with the functioning of the muscle cells. Carnitine deficiency, which also has been reported in humans, produces symptoms similar to CAT-I deficiency. A deficiency of the translocase would be expected to have similar biochemical and clinical effects.

2.2. β OXIDATION OF FATTY ACIDS

Fatty acids represent an important source of energy for the body. This energy is released when the fatty acids are oxidized. When fatty acids are activated to CoA derivatives, they are transported into the matrix of the mitochondria via the carnitine cycle and undergo β oxidation. Four reactions are

$$(CH_3)_3N\text{-}CH_2\text{-}\underset{\underset{OH}{|}}{CH}\text{-}CH_2\text{-}COOH$$

Figure 2-4. Structure of carnitine.

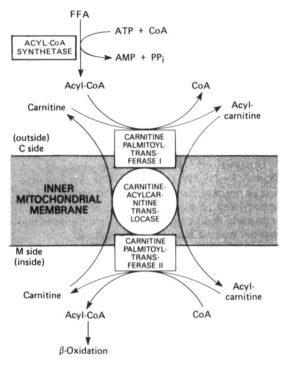

Figure 2-5. Carnitine cycle. (From Murray et al., 1988. Reproduced with permission.) CAT-I, Carnitine palmitoyl-transférase I; CAT-II, carnitine palmitoyl-transferase II; FFA, free fatty acids.

involved in this process. In the first reaction, dehydrogenation produces a 2,3-*trans* unsaturated derivative; in the second reaction, hydration of the double bond forms a 3-hydroxy derivative; in the third reaction, dehydrogenation yields a β-keto derivative; and in the fourth reaction, a thiolytic cleavage by CoA yields acetyl-CoA and a fatty acyl-CoA having two fewer carbon atoms. Successive repetitions of these reactions result in the conversion of even-numbered carbon atom fatty acids to acetyl-CoA. Fatty acids containing an odd number of carbon atoms yield several molecules of acetyl-CoA and one molecule of propionyl-CoA per molecule of fatty acid. The overall biochemical equation for oxidation of the saturated 16 carbon palmitic acid is shown in Fig. 2-6.

$$Palmityl\text{-}CoA + 7FAD + 7NAD^+ + 7CoA + 7H_2O \text{-------------->}$$
$$8acetyl\text{-}CoA + 7FADH_2 + 7NADH + 7H^+$$

Figure 2-6. Overall oxidation of palmitic acid.

The oxidation of unsaturated fatty acids also occurs in the mitochondria but requires three additional enzymes: an isomerase, a reductase, and an epimerase. The isomerase is required to convert a δ-3-*cis* double bond to a δ-2-*trans* double bond. The epimerase is required to convert a hydroxyl group in the D configuration to one in the L configuration. The reductase converts the *trans* 2,3-*cis*-4,5-diene intermediate to the *trans* 3,4 intermediate (Schulz and Kunau, 1987). According to Schulz and Kunau, the epimerase-dependent pathway is inoperative in mitochondria but may play a minor role in nonmitochondrial β-oxidation systems. The reductase-dependent pathway is the major pathway in mitochondria.

The β oxidation of fatty acids yields the reduced coenzymes $FADH_2$ and NADH. The four hydrogen atoms that are transferred from the fatty acid to the reduced coenzymes are oxidized via the electron transport chain in the inner mitochondrial membrane. The oxidation of 1 mole of $FADH_2$ yields 2 moles of ATP and the oxidation of 1 mole of NADH yields 3 moles of ATP. Thus, each turn of the β-oxidation cycle yields 5 moles of ATP per mole of saturated fatty acid and requires 1 mole of O_2.

The oxidation of each mole of acetyl-S-CoA via the tricarboxylic acid cycle in the mitochondria yields 12 moles of ATP. Therefore, a total of 131 moles of ATP is produced from the oxidation of 1 mole of palmitic acid, which requires seven turns of the β-oxidation cycle and yields 7 $FADH_2$, 7 NADH, and 8 acetyl-S-CoA. However, two equivalents of ATP are needed to activate the free palmitic acid by the ATP-requiring kinase since it is assumed that the pyrophosphate produced in the activation reaction is hydrolyzed to orthophosphate. The net yield of ATP is 129. Assuming a standard free energy of hydrolysis of 1 mole of ATP under standard conditions to be -7.3 kcal, this represents a conservation of about 942 kcal of chemical energy as ATP. This energy yield is approximately 40% of the 2348 kcal released when 1 mole (256 g) of palmitic acid is oxidized to CO_2 and H_2O in a bomb calorimeter.

The oxidation of even-chain unsaturated fatty acids yields less ATP than does oxidation of the corresponding saturated even-chain fatty acids, since the presence of each double bond results in one less $FADH_2$ molecule and hence two less ATPs.

2.3. DISORDERS OF PROPIONIC ACID METABOLISM

The oxidation of odd-chain fatty acids must take into account the formation of 1 mole of propionic acid per mole of fatty acid. Odd-chain fatty acids occur in small amounts in most foods consumed by humans. Propionic acid is also produced by the metabolism of certain amino acids such as valine, isoleucine, methionine, and threonine. The metabolism of propionic acid is shown Fig. 2-7.

$$\text{Propionyl-CoA} + \text{ATP} + \text{CO}_2 \xrightarrow{\text{carboxylase}} \text{D-methylmalonyl-CoA} + \text{ADP} + \text{Pi}$$

$$\text{D-Methylmalonyl-CoA} \xrightarrow{\text{racemase}} \text{L-methylmalonyl-CoA}$$

$$\text{L-Methylmalonyl-CoA} \xrightarrow{\text{mutase-vitamin B12}} \text{succinyl-CoA}$$

Figure 2-7. Metabolism of propionic acid.

2.3.1. Propionyl-CoA Carboxylase Deficiency

Genetic defects in the metabolism of propionic acid occur in humans. A deficiency in carboxylase, which has been reported in more than 30 children, gives rise to propionic acid aciduria. In this condition, there is an accumulation of propionic acid, which lowers the pH of the blood and leads to excessive excretion of propionic acid in the urine. Recurrent attacks of ketoacidosis and hyperammonemia are common problems in this disease. These symptoms are aggravated by high protein intake. The range of clinical expression varies from death in early infancy to a relatively mild course with developing mental retardation. Dietary amino acids such as isoleucine, threonine, valine, and methionine induce the ketosis. Since the carboxylase is a biotin-dependent enzyme, it is expected that a deficiency of biotin will produce some of the same symptoms as carboxylase deficiency. However, biotin is a cofactor for other carboxylases such as acetyl-CoA carboxylase and pyruvate carboxylase, and hence the biochemical disorder induced by biotin deficiency can be more complex. The biotin-responsive form of propionyl-CoA carboxylase deficiency indicates that in this abnormality there is an inability of biotin to be incorporated into the enzyme. Large doses of biotin can in part rectify the latter defect. If the mutation in the gene for the enzyme is such that no mRNA is made, or the enzyme is so altered that it cannot function with or without biotin, the biotin-unresponsive form of the disease results.

2.3.2. Methylmalonic Acid Acidemias

At least five inherited forms of methylmalonic acid acidemias have been found in humans and more than 50 cases have been reported. A deficiency of the racemase gives rise to D-methylmalonic acid aciduria in which large amounts of

D-methylmalonic and propionic acids are excreted in the urine. A deficiency of the mutase is associated with accumulation of both D- and L-methylmalonic acid and possibly propionic acid. Persons who are deficient in vitamin B12 become anemic and often excrete excessive amounts of propionic acid and methylmalonic acid in the urine. This is due to the inactivation of the mutase, which requires vitamin B12 as a cofactor. Defects in cobalamin metabolism also are involved in the methylmalonic acid acidemias.

2.4. α AND ω OXIDATION OF FATTY ACIDS

Although β oxidation is the major way in which fatty acids are oxidized, some oxidation occurs via α and ω oxidation. α oxidation converts the fatty acid to a 2-hydroxy derivative, which is decarboxylated to a fatty acid having one less carbon atom. The initial hydroxylation is catalyzed by a mitochondrial monooxygenase, which requires O_2, Mg^{2+}, and NADPH. The decarboxylation step appears to occur in the endoplasmic reticulum and requires O_2, Fe^{2+}, and vitamin C.

ω oxidation occurs at the methyl-terminal end of the fatty acid and proceeds similarly to β oxidation. It is a minor pathway for the oxidation of fatty acids.

2.5. REFSUM'S DISEASE

Dairy products and ruminant fats appear to be the major dietary source of phytanic acid. Phytanic acid oxidase is responsible for the oxidation of phytanic acid (3,7,11,15-tetramethylhexadecanoic acid). The first step, which involves hydroxylation at the α position of phytanic acid, is the critical step that is defective. Since phytanic acid cannot be oxidized by β oxidation in mitochondria, large amounts of phytanic acid accumulate in tissues and serum of humans with Refsum's disease. This is an inheritable disorder of the nervous system in which phytanic acid oxidase is either missing or defective. It is believed that the bulky tetramethyl fatty acids disturb the structure and fluidity of myelin, which in turn upsets the normal function of the proteins in this membrane and leads to brain damage. Persons with Refsum's disease must avoid foods rich in phytanic acid.

2.6. ZELLWEGER'S SYNDROME

Zellweger's syndrome is a rare familial cerebro-hepato-renal condition affecting many tissues and is lethal within the first year after birth. In this disease liver and kidneys lack peroxisomes but mitochondria appear abnormal. Whether peroxisomes are completely absent remains to be determined. Defects in bile acid biosynthesis also are present.

Peroxisomes are small cellular particles containing urate oxidase, D-amino acid oxidase, α-hydroxyacid oxidase, dihydroxyacetone phosphate (DHAP) acyl transferase, alkyl-DHAP synthase, DHAP-NADPH oxidoreductase, glycerol-phosphate dehydrogenase, isocitrate dehydrogenase, and catalase. Catalase oxidizes hydrogen peroxide to molecular oxygen and water. Some of the hydrogen peroxide is produced by oxidases, which carry out a β-type oxidation of fatty acids. Unlike β oxidation in mitochondria which traps the energy as ATP, β oxidation in peroxisomes generates energy as heat. In Zellweger's syndrome protein transport also is impaired. Patients with this disease also have impaired peroxisomal function and impaired ability to synthesize ether-type lipids (alkenyl and alkanyl lipids). It has been postulated that if the key enzymes for ether-type lipid biosynthesis were restricted to peroxisomes, the peroxisome would be indispensable for normal mammalian development.

2.7. KETOGENESIS, KETOSIS, AND KETOACIDOSIS

Under normal conditions, the oxidation of fatty acids in liver is regulated so that excess acetyl-CoA production does not occur. However, when fatty acids are presented to the liver faster than they can be oxidized to CO_2 and H_2O, a surplus of acetyl-CoA results. The excess acetyl-S-CoA in the mitochondria is converted to β-hydroxy-β-methylglutaryl-CoA (HMG-CoA) which is cleaved to form acetyl-CoA and free acetoacetic acid. Some of the free acetoacetic acid is reduced in the mitochondria to β-hydroxybutyric acid and a small amount is spontaneously decarboxylated to acetone. Acetoacetic acid, β-hydroxybutyric acid, and acetone are known as ketone bodies. Normal production of these ketone bodies is called ketogenesis. Abnormal production of excessive amounts is referred to as ketosis. The reactions leading to the production of ketone bodies are shown in Fig. 2-8.

The liver has a very limited ability to oxidize free acetoacetic acid, since it lacks the two enzymes, acetoacetyl-CoA synthetase and succinyl-CoA thiophorase, necessary to convert acetoacetic acid back to acetoacetyl-CoA. Acetoacetic acid therefore must leave the liver, enter the blood, and go to extrahepatic tissues for oxidation. The major tissues that can oxidize acetoacetic acid are skeletal muscle, heart muscle, kidney, and brain since they contain the enzymes that can convert acetoacetic acid to acetoacetyl-CoA.

As shown in Fig. 2-9, acetyl-CoA can be produced from the degradation of certain ketogenic amino acids, from glycolysis, and from β oxidation of fatty acids. When the production of acetyl-CoA exceeds the capacity of the liver

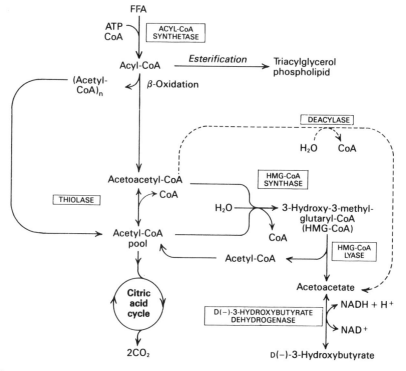

Figure 2-8. Reactions leading to the synthesis of ketone bodies in the liver. FFA, free fatty acids. (From Murray et al., 1988. Reproduced with permission.)

mitochondria to oxidize it to CO_2 and H_2O, then acetyl-CoA is shunted into acetoacetyl-CoA, which is converted into HMG-CoA, the main intermediate in ketogenesis. Excess ketogenesis occurs during prolonged starvation and in uncontrolled diabetes. During starvation the brain adapts its metabolic system to use ketone bodies and thereby spares glucose. Muscle also utilizes ketone bodies and fatty acids for energy during starvation, which helps prevent hypoglycemia by sparing glucose. During this time gluconeogenesis is stimulated in order to maintain blood glucose levels near normal.

The substrate and hormone levels in the blood plasma of well-fed, fasting, and starving humans are shown in Table 2-3. It can be seen that the insulin/glucagon ratio is high in the fed state and falls markedly after a 3-day fast but remains constant up to a 5-week fast. Glucose levels fall modestly 12 hr after feeding but are maintained at a level of 3.6 mM, even after 5 weeks of starvation, mainly as a result of gluconeogenesis. The levels of free fatty acids, acetoacetate, and β-hydroxybutyrate increase markedly after a 3-day fast and remain

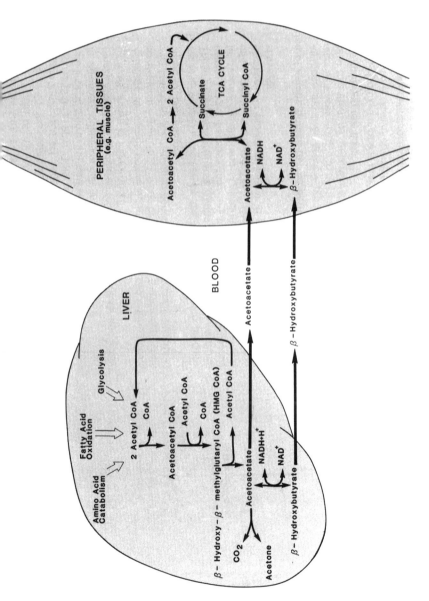

Figure 2-9. Formation of ketone bodies in liver and their utilization by nonhepatic cells. (From Champe and Harvey, 1987. Reproduced with permission.)

Table 2-3. Plasma Hormone and Substrate Changes during Fasting[a]

Hormone/substrate levels	Well fed	12 hr	Fasting 3 days	5 weeks
Insulin (μU/ml)	40	15	8	6
Glucagon (pg/ml)	80	100	150	120
Insulin/glucagon ratio	0.5	0.15	0.05	0.05
Glucose (mM)	6.1	4.8	3.8	3.6
Fatty acids (mM)	0.14	0.6	1.2	1.4
Acetoacetate (mM)	0.04	0.05	0.4	1.3
β-Hydroxybutyrate (mM)	0.03	0.10	1.4	6.0

[a]From Ruderman et al., 1976. Reproduced with permission.
[b]Data are for normal-weight subjects except for the 5-week starvation values, which are from obese subjects undergoing therapeutic starvation.

elevated during the prolonged starvation. The increase in free fatty acids results from the stimulation of lipolysis in adipose tissue by hormones such as glucagon, epinephrine, and ACTH, which are elevated during starvation. These hormones exert their lipolytic effects more strongly because of the low levels of insulin, which is the major lipogenic and antilipolytic hormone. There is also increased urinary excretion of acetoacetic acid and β-hydroxybutyric acid during starvation. Urinary excretion of acetoacetic acid increases from 0.05 mmole/day to 11 mmole/day after an 8-day fast, whereas β-hydroxybutyrate excretion increases from 0.03 mmole/day to 77 mmole/day.

The presence of elevated levels of ketone bodies in the blood is called ketonemia; elevation of the levels of these ketones in the urine is called ketonuria. Since excretion of these acids in the urine requires that they be neutralized by Na^+ and K^+, their loss in the urine leads to a loss of these cations in the body. The excessive production of acetoacetic acid and β-hydroxybutyric acid causes the blood pH to drop from a normal value of 7.4 to as low as 6.8. This lowering of blood pH is named acidosis. When it is due to ketosis, the condition is known as ketoacidosis. Ketoacidosis, a potentially life-threatening condition, is a serious problem in uncontrolled diabetes and in severe starvation.

The changes in the levels of plasma ketones to plasma glucose and fatty acids in humans are shown in the Fig. 2-10. It is apparent that plasma ketone levels change more dramatically than do plasma glucose and fatty acid levels, especially after the second day of fasting. Since the fatty acids released by lipolysis are taken up by several tissues, and since 1 mole of fatty acid (16 or 18 carbon atoms) yields 4 to 4.5 moles of acetoacetate, the level of fatty acid in

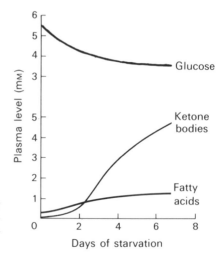

Figure 2-10. Changes in levels of plasma fatty acids, ketone bodies, and glucose during starvation. (From Stryer, 1988. Reprinted with permission.)

plasma reaches a plateau, whereas the level of ketone bodies continues to rise. The slowing down of the fall in plasma glucose is due to enhanced gluconeogenesis, which is mediated by elevated levels of fatty acids and acetyl-CoA in the liver.

2.8. SYNTHESIS OF FATTY ACIDS

The two major types of fatty acids in humans are saturated and unsaturated even-chain fatty acids containing 16–24 carbon atoms. Saturated fatty acids have no double bonds, in contrast to unsaturated fatty acids, which contain two to six double bonds. Fatty acids containing two or more double bonds are referred to as polyunsaturated. The structures of these fatty acids are given in Chapter 1 (Fig. 1-1). The human body can synthesize palmitic, stearic, oleic, eicosapentenoic, and docosahexenoic acids, but not linoleic and linolenic acids. The latter two acids are therefore essential dietary constituents for humans just as are the essential amino acids.

Fatty acids are synthesized from acetyl-CoA. The liver, adipose tissue, and mammary gland are the major organs for this synthesis, and the hormone insulin plays an important role in stimulating the synthesis of fatty acids. Insulin stimulates the synthesis of a variety of lipogenic enzymes, including acetyl-CoA carboxylase, citrate lyase, malic enzyme, malate dehydrogenase, glucose-6-phosphate dehydrogenase, and the fatty acid synthase (FAS) complex. This process is called long-term or chronic regulation of lipogenesis. Insulin also

$$\text{Acetyl-CoA + ATP + HCO}_3^- \xrightarrow{\text{acetyl-CoA carboxylase}} \text{malonyl-CoA + ADP + Pi}$$

$$\text{Acetyl-CoA + 7 malonyl-CoA + 14NADPH + 7H}^+ \xrightarrow{\text{FAS}}$$

$$\text{palmitic acid + 7CO}_2 + 14\ \text{NADP}^+ + 8\text{CoA + 6H}_2\text{O}$$

Figure 2-11. Conversion of acetyl-CoA to palmitic acid.

exerts a short-term (acute) regulation of lipogenesis by enhancing the dephosphorylation of acetyl-CoA carboxylase, which is the rate- limiting enzyme in fatty acid synthesis. The dephospho form of acetyl-CoA carboxylase is the active form of the enzyme.

The major fatty acid made from acetyl-CoA is palmitic acid. The synthesis involves a reductive condensation process utilizing the coenzyme NADPH and is catalyzed by the FAS multienzyme complex (Fig. 2-11).

Acetyl-CoA carboxylase of *Escherichia coli* consists of three integrated enzymes: biotin carboxyl carrier protein (BCP; mol. wt. 45,000), which has two identical subunits, with biotin linked to the ε-amino group of lysine; biotin carboxylase (BC; mol. wt. 98,000), which has two identical subunits; and trans-carboxylase (TC; mol. wt. 130,000), which has two pairs of subunits, mol. wt. 35,000 and 30,000. The partial reactions of acetyl-CoA carboxylase are given in Fig. 2-12.

The FAS complex of chicken liver (mol. wt. 500,000) has six α and six β chains. There are six active sites on the enzyme complex, but only four appear to be functional at any time during synthesis of the fatty acid. The major product is the 16-carbon palmitic acid. The elongation stops at 16-carbon chain length, mainly because the palmityl group on the acyl carrier protein (ACP) is too bulky to be transferred rapidly enough to the cysteine-SH group of the FAS. This allows the thioesterase to hydrolyze off the free palmitic acid and release CoA,

$$\text{BCP + HCO}_3^- + \text{ATP} \xrightarrow{\text{BC}} \text{BCP-COO}^- + \text{ADP + Pi}$$

$$\text{BCP-COO}^- + \text{acetyl-CoA} \xrightarrow{\text{TC}} \text{malonyl-CoA + BCP}$$

Figure 2-12. Conversion of acetyl-CoA to malonyl-CoA. This is the rate-limiting regulatory step in the synthesis of fatty acids.

which is essential for the cycle to operate.

Foods that can be metabolized to yield acetyl-CoA can be converted to fatty acids. The main dietary source, which yields the acetyl-CoA in humans, is carbohydrate. Excess dietary carbohydrate is stored as glycogen and as fatty acids (the latter in the form of triacylglycerols).

2.8.1. Elongation of Fatty Acids

The FAS complex produces primarily palmitic acid. To make the longer-chain fatty acids, two elongation systems present in liver and other cells convert palmitic acid (C16) to stearic (C18), arachidic (C20), behenic (C22), and lignoceric (C24) acids. Thus the C16 palmitic acid is elongated to linear-chain fatty acids, having 18–24 carbon atoms. The mitochondrial elongation enzyme complex does this by the sequential addition of two carbon atoms, utilizing acetyl-CoA, NADH, and NADPH.

The microsomal elongation system uses both saturated and unsaturated fatty acyl-CoA derivatives and elongates these by using malonyl-CoA. The intermediates are similar to those in the fatty acid synthase reactions except that the intermediates are not linked to ACP. In humans, however, palmitic and palmitoleic acids cannot be elongated to linoleic and linolenic acids.

2.8.2. Formation of Nonessential Monoenoic and Polyenoic Fatty Acids

Animals and humans can convert palmitic acid to palmitoleic acid and can convert stearic acid to oleic acid by introducing a double bond in the corresponding saturated fatty acid. This dehydrogenation is carried out by the endoplasmic reticulum stearyl desaturase. This enzyme is a mixed-function oxygenase that uses molecular O_2, NADH, FAD, NADH-cytochrome b_5 reductase, cytochrome b_5, and a desaturase. The reactions involve the abstraction of hydrogen, the addition of a hydroxyl group, and the elimination of water to produce a double bond at the 9,10 position. First, electrons are transferred from NADH to FAD of the NADH-cytochrome b_5 reductase. The heme iron atom of cytochrome b_5 is reduced to the ferrous form. The nonheme iron atom of the desaturase is converted to the Fe^{2+} state, which enables it to react with molecular O_2 and the saturated fatty acyl chain of stearyl-CoA. A double bond is formed, and two molecules of water are released. Two electrons come from NADH and two come from the single bond of the fatty acyl chain during the abstraction of two hydrogen atoms. Insulin stimulates the synthesis of the desaturase.

Animal and human cells contain a variety of polyunsaturated fatty acids, some derived from the diet and some made de novo. The C18 polyunsaturated fatty acids made de novo all have double bonds between carbon atoms 1 and 9, beginning with the carboxyl group. These fatty acids can be made by alternate desaturation and chain elongation, starting with oleic acid. However, polyunsaturated fatty acids, in which one or more double bonds are located within the last seven carbon atoms of the fatty acid, cannot be made de novo. These are the essential dietary fatty acids, i.e., linoleic and linolenic acids. However, once these essential fatty acids are taken in the diet, they can be elongated and desaturated to yield other polyunsaturated fatty acids in which double bonds are made between carbon atoms 1 and 7. Why living cells need such an array of different fatty acids is not fully understood. They are required in certain combinations, esterified to phospholipids, to regulate the fluidity of cell membranes. The fluidity of the cell membrane is important in regulating the function of membrane proteins. Palmitic acid (C16) can be elongated to stearic acid (C18), which may be further elongated to C20–C24 fatty acids. Stearic acid can also be desaturated to oleic acid, which can then be converted to eicosatrienoic acid and nervonic acid. Lignoceric acid and nervonic acid are important constituents of brain cells.

2.9. ESSENTIAL FATTY ACID DEFICIENCY

Two dietary fatty acids, linoleic and linolenic acids, are essential for human life, since they cannot be synthesized by the body. Arachidonic acid, formed from linoleic acid, is not essential in humans if enough linoleic acid is taken in the diet. A deficiency of the essential fatty acids leads to skin lesions, fragile red blood cells, loss of hair, weight loss, kidney damage, sterility, and possibly death. The intake of 1–2% of the total dietary energy requirement as linoleic and linolenic acids is sufficient to prevent essential fatty acid deficiency.

Linoleic acid and linolenic acid are essential in part because they are precursors for the prostaglandins, leukotrienes, and lipoxins. These essential fatty acids also help maintain a critical fluidity state in cell membranes. However, they may have other functions yet to be discovered. Linoleic acid has been found to occur in high amounts in epidermal acylglucosylceramides and acylceramides.

Acylglycosylceramides and acylceramides are sphingolipids that act as barriers to water permeability and prevent water loss in the skin. The sphingolipids are formed in lamellar bodies of the stratum granulosum and are secreted into the interstices during differentiation of these cells. They function as a barrier to water loss in the stratum lucidum. The lipids of the epidermal water barrier also contain sterol esters, sterols, ceramides, triacylglycerols, and long-chain saturated fatty acids having 22–26 carbon atoms. During epidermal terminal differentiation, lipids undergo changes in both composition and localization. A mixture of polar and neutral lipids is replaced by a more nonpolar mixture of lipids.

These lipids are organized in lamellar bilayers in the stratum corneum interstices, where they are believed to provide a barrier against water loss necessary for terrestrial life. Grubauer et al. (1989a,b) have shown that removal of nonpolar lipids alone from the stratum corneum causes only a modest level of barrier disruption, whereas removal of sphingolipids and free sterols leads to a more profound level of barrier perturbation. These workers also report that trans-epidermal water flux is the signal that regulates epidermal lipid synthesis, which is associated first with the redeposition of stratum corneum lipids and then with normalization of stratum corneum barrier function.

The essential fatty acid linolenic acid can be converted to docosahexenoic acid (DHA), which occurs in high amounts in the retina and gray matter of the cerebral cortex. DHA is necessary for normal retinal function.

Polyunsaturated fatty acids are important intermediates for the syntheses of prostaglandins, leukotrienes, and lipoxins. These are hormonelike compounds that influence the function of most cells and play an important role in inflammation and in the activities of the kidney, heart, stomach, lung, brain, uterus, and probably most organs of the body. The roles of prostaglandins in platelet aggregation and atherogenesis are discussed in Chapters 7 and 8.

Polyunsaturated fatty acids also regulate the fluidity of cell membranes, which is important in the functioning of membrane proteins and for the permeability properties of membranes. Apparently, very subtle changes in the fatty acid composition of cell membranes can have dramatic effects on the functions of certain membrane proteins.

REFERENCES

Borst, P., 1983, Animal peroxisomes (microbodies), lipid biosynthesis and the Zellweger syndrome, *Trends Biochem. Sci.*, 8:269.
Borum, R., 1981, Possible carnitine requirement of the newborn and the effect of genetic disease on the carnitine requirement, *Nutr. Rev.*, 39:385.
Cahill, G. F., Jr., 1976, Starvation in man, *Clin. Endocrinol. Metabol.*, 5:398.
Carnitine metabolism in man, 1980, *Nutr. Rev.*, 38:338.
Champe, P. C., and Harvey, R. A., 1987, *Lippincott's Illustrated Reviews: Biochemistry*, J. B. Lippincott Co., New York.
Devlin, T. M. (ed.), 1982, *Textbook of Biochemistry with Clinical Correlations*, John Wiley & Sons, New York.
DiMauro, S., and Melis DiMauro, P. M., 1973, Muscle carnitine palmityltransferase deficiency and myoglobinuria, *Science*, 182:929.
Engel, A. G., and Angelini, C., 1973, Carnitine deficiency of human skeletal muscle with associated lipid storage myopathy: a new syndrome. *Science*, 179:899.
Essential fatty acids and maintenance of the epidermal water barrier, 1986, *Nutr. Rev.*, 44:151.
Fatty acid binding protein in heart energy production, 1985, *Nutr. Rev.*, 43:348.
Goodridge, A. G., 1986, Regulation of the gene for fatty acid synthase, *Fed. Proc.*, 45:2399.

Grubauer, G., Feingold, K. R., Harris, R. M., and Elias, P. M., 1989a, Lipid content and lipid type as determinants of epidermal permeability barrier, *J. Lipid Res.*, 30:89.

Grubauer, G., Elias, P. M., and Feingold, K. R., 1989b, Transepidermal water loss: the signal for recovery of barrier structure and function, *J. Lipid Res.*, 30:323.

Hoppel, C. L., 1982, Carnitine and carnitine palmitoyltransferase in fatty acid oxidation and ketosis. *Fed. Proc.*, 41:2853.

Linoleic acid deficiency in man, 1982, *Nutr. Rev.,*40:144.

Mattick, J. S., Nickless, J. Mizugaki, M., Yang, C., Uchiyama, S., and Wakil, S. J., 1983a, The architecture of the animal fatty acid synthetase, part II, *J. Biol. Chem.*, 258:15300.

Mattick, J. S., Tsukamoto, Y., Nickless, J., and Wakil, S. J., 1983b, The architecture of the animal fatty acid synthetase, part I, *J. Biol. Chem.*, 258:15291.

McGarry, J. D., and Foster, D. W., 1979, In support of the roles of malonyl-CoA and carnitine acyltransferase I in the regulation of hepatic fatty acid oxidation and ketogenesis, *J. Biol. Chem.*, 254:8163.

McGarry, J. D., Leatherman, G. F., and Foster, D. W., 1978, Carnitine palmitoyltransferase I: The site of inhibition of hepatic fatty acid oxidation by malonyl-CoA, *J. Biol. Chem.*, 253:4128.

McGilvery, R. W., and Goldstein, G. W., 1983, *Biochemistry, A Functional Approach,* W. B. Saunders Co., New York.

Murray, R. K., Granner, D. K., Mayes, P. A., and Rodwell, V. W., 1988, *Harper's Biochemistry,* 21st ed., Appleton & Lange, Publ. Norwalk, Connecticut.

Neuringer, M., and Connor, W. E., 1986, Omega-3 fatty acids in the brain and retina: evidence for their essentiality, *Nutr. Rev.*, 44:285.

Press, M, Kikuchi, H., Shimoyama, T., and Thompson, G. R., 1974, Diagnosis and treatment of essential fatty acid deficiency in man, *Brit. Med. J.*, 2:247.

Ruderman, N. B., Aoki, T. T., and Cahill, G.F., Jr., 1976, Gluconeogenesis and its disorders in man, in: Gluconeogenesis, Its Regulation in Mammalian Species (R. W. Hanson and M. A. Mehlman, eds., p. 515, John Wiley & Sons, New York.

Schulz, H., and Kunau, W., 1987, Beta-oxidation of unsaturated fatty acids: a revised pathway, *Trends Biochem. Sci.*, 12:403.

Smith, E. L., Hill, R. L., Lehman, I. R., Lefkowitz, R. J., Handler, P, and White, A., 1983a, *Principles of Biochemistry: Mammalian Biochemistry,* 7th ed., McGraw-Hill Book Co., New York.

Smith, E. L., Hill, R. L., Lehman, I. R., Lefkowitz, R. J., Handler, P., and White, A., 1983b, *Principles of Biochemistry: General Aspects,* 7th ed., McGraw-Hill Book Co., New York.

Stryer, L., 1988, *Biochemistry,* 3rd ed., W. H. Freeman & Co., San Francisco.

Systemic carnitine deficiency, 1981, *Nutr. Rev.*, 39:400.

Tsukamoto, Y., Wong, H., Mattick, J. S., and Wakil, S. J., 1983, The architecture of the animal fatty acid synthetase, part IV, *J. Biol. Chem.*, 258:15312.

Unterberg, C., Heidl, G., von Basseqitz, D. B., and Spener, F., 1986, Isolation and characterization of the fatty acid binding protein from human heart, *J. Lipid Res.*, 27:1287.

Wong, W., Mattick, J. S., and Wakil, S. J., 1983, The architecture of the animal fatty acid synthetase, part III, *J. Biol. Chem.*, 258:15305.

Chapter 3

DISORDERS OF EXCESSIVE ALCOHOL INTAKE
Hypoglycemia, Fatty Liver, and Liver Cirrhosis

3.1. INTRODUCTION TO THE BIOCHEMICAL EFFECTS OF ALCOHOL INTAKE

Ethanol (CH_3CH_2OH), also called ethyl alcohol or simply alcohol by most biologists and clinicians, is a primary alcohol. It is soluble both in water and in lipid solvents. Indeed, ethanol is commonly used to extract lipids from tissues. The major industrial source of ethanol is fermentation of glucose or carbohydrates, which can be broken down to glucose. Very little ethanol is produced by metabolic processes in animal cells. The oxidation of 1 g of ethanol in the body to carbon dioxide and water yields 7 kcal of energy and produces 0.9 g of water. In this chapter the metabolism of ethanol (which will also be referred to as alcohol) and the biochemical and clinical effects of excessive intake of alcohol will be considered.

It is estimated that the average adult American consumes about 3 gallons of alcohol per year and that approximately 10 million Americans are alcoholics. About 3 million teenagers in the United States drink in excess. Alcohol-related deaths in the United States number close to 150,000 per year.

Alcoholic liver cirrhosis ranks as the sixth most common cause of death in the United States and close to the third most common in New York City. Alcohol abuse costs the United States about $40 billion per year due to loss of work, health costs, motor accidents, and fire loss. Fifty percent of highway fatalities are attributed to excessive alcohol consumption.

Alcohol affects all organs of the body. Excessive alcohol intake inhibits protein synthesis in liver, heart, and muscle and can cause heart enlargement, muscle weakness, fatty liver, alcoholic hepatitis, liver cirrhosis, hepatic coma, hypoglycemia, ketoacidosis, lactacidosis, hyperuricemia, portal hypertension, ascites (fluid accumulation in the abdomen), and esophageal varices, which can rupture and cause death by uncontrolled bleeding. Failure of the liver to produce clotting proteins aggravates the bleeding. Excessive intake of alcohol also leads to malnutrition and vitamin deficiencies. A deficiency of vitamin B12 and folic acid can lead to anemia. Alcohol intake inhibits fetal growth, especially in the last trimester, and can lead to brain damage. This abnormality is known as fetal alcohol syndrome.

The harmful effects of excessive alcohol intake result in part from an inhibition of ATPases, inhibition of calcium transport into or out of cells, inhibition of protein synthesis, and the fluidizing effect of alcohol on cell membranes. The harmful biochemical effects include mitochondrial damage, inhibition of fatty acid oxidation, inhibition of gluconeogenesis, hypoglycemia, and increased synthesis of VLDL in liver. The latter increases the level of triacylglycerols (triglycerides) in plasma. Many of the biochemical effects of alcohol intake are attributed to the high NADH/NAD+ ratio that prevails when alcohol is oxidized to acetaldehyde and then to acetate.

3.2. METABOLISM OF ALCOHOL

Alcohol is metabolized primarily in the liver (Fig. 3-1). The first step is mediated by the cytosol enzyme alcohol dehydrogenase (ADH) and requires the coenzyme NAD+. This reaction is believed to be the rate-limiting step and is responsible for the zero order kinetics of ethanol disappearance from the blood. The second reaction, catalyzed by aldehyde dehydrogenase (ALDH), also requires NAD+. Thus, the oxidation of 1 mole of ethanol to acetate produces 2 moles of NADH, and this increases very markedly the NADH/NAD+ ratio in the liver cell. The acetate is converted to acetyl-CoA and oxidized to CO_2 and H_2O in the mitochondria. Some of the acetyl-CoA is converted to fatty acids and to ketone bodies.

Both ADH and ALDH occur as different genetically determined isoenzyme

$$\underset{\text{Ethanol}}{CH_3CH_2OH} + NAD^+ \xrightarrow{\quad ADH \quad} \underset{\text{acetaldehyde}}{CH_3CHO} + NADH + H^+$$

$$\underset{\text{Acetaldehyde}}{CH_3CHO} + NAD^+ \xrightarrow{\quad ALDH \quad} \underset{\text{acetate}}{CH_3COOH} + NADH + H^+$$

Figure 3-1. Oxidation of ethanol in liver.

Table 3-1. Isoenzymes of ADH[a]

Enzyme class	Gene locus	Peptide chains	K_m (mM)
I	ADH-1	α	
	ADH-2	β1	
		β2 (Berne)	>5
		β (Honolulu)	
		β (Indianapolis)	
	ADH-3	γ1 γ2	
II	?	π	34
III	?	λ	Very high

[a]From von Wartburg and Buchler, 1984. Reproduced with permission.

forms. ADH consists of three chains (α, β, and γ), and has several different isoenzyme species (Table 3-1).

Many multiple forms are ascribed to the presence of isoenzymes. These isoenzymes occur by random combination of three different subunits to produce six possible dimeric forms. Additional enzyme forms are due to enzyme polymorphisms. At the gene locus coding for the γ-polypeptide chains, there are allelic genes that lead to the γ1 and γ2 subunits. Apparently, several polymorphisms occur at the gene locus that codes for the β subunits. In addition to the normal β1 subunit, the atypical β2 and β-Indianapolis forms occur. The π-ADH form is present in all humans to variable extents.

The K_m for ethanol for the various forms of human liver ADH varies widely. The main isoenzyme, ββ1, isolated from normal liver, reveals nonlinear kinetics, with a lower K_m of 0.5 mM and a higher K_m of about 8 mM. The π enzyme has a high K_m of 15–30 mM.

ALDH occurs both in the cytoplasm and mitochondria. Very little is known about human ALDH and its genetic control. The cytosol enzyme in rats has a high K_m for alcohol and the mitochondrial enzyme has a low K_m. In humans, in contrast, a significant fraction of acetaldehyde oxidation is localized in the cytoplasm rather than primarily in the mitochondria. The cytosol and mitochondrial ALDHs have similar K_ms for acetaldehyde, but the cytosol enzyme can also use $NADP^+$ as a cofactor and is more sensitive to inhibition by disulfiram (Antabuse). These findings indicate that aldehyde toxicity in humans may not be due primarily to specific damage to mitochondria, as prevails in rats.

A large number of Orientals lack the low-K_m form of ALDH and have a very low tolerance for alcohol because of the buildup of acetaldehyde in the blood and its effect on the central nervous system (CNS) and heart.

ADH is found in many tissues and often in specialized cells within a particular tissue. The large variety of enzyme phenotypes of both ADH and ALDH may account in part for individual differences in the metabolism of alcohol and may play a role in the development of alcoholism.

ADH is inhibited by pyrazole whereas ALDH is inhibited by disulfiram. Disulfiram was originally synthesized by a Danish chemist as one of a series of anthelminthic compounds. After spending the day working with the material, the chemist went to a cocktail party, where he became violently ill after a single drink of alcohol. This unpleasant and potentially dangerous reaction most likely results from the accumulation of acetaldehyde. Disulfiram is used for the treatment of alcoholics. The clinical effects of the drug include severe flushing, throbbing headache, nausea, sweating, vertigo, chest pain, and confusion. These severe reactions often persuade alcoholics to stop drinking alcohol.

Another pathway for alcohol oxidation is mediated by a liver mixed-function microsomal enzyme oxidizing system named MEOS. This NADPH-requiring enzyme system, which has a K_m of 8–10 for ethanol, is believed to be more important at higher levels of cellular alcohol. Acute alcohol consumption leads to an inhibition of hepatic MEOS, whereas chronic intake leads to an increase in this enzyme system as a result of hyperplasia of the endoplasmic reticulum.

When alcohol is consumed over a short period of time, the blood alcohol concentration (BAC) is approximately a linear function of the amount ingested. One 12-oz bottle of beer is the equivalent of 1 oz of 100-proof bourbon, or about 15 ml of absolute alcohol. The liver requires about 1–1.5 hr to oxidize 1 oz of alcohol (equivalent to one shot of whiskey, 12 oz of 3% beer, or 4 oz of 10% wine). One ounce of alcohol will produce a BAC of approximately 0.02%–0.03% (20–30 mg/dl). In the United States, persons who have a BAC of 100 mg/dl or higher are legally intoxicated. (A concentration of 100 mg/dl, equivalent to 22 mM, is sometimes designated as 0.1%.) A person who ingests five 12-oz bottles of beer or five shots of whiskey over a period of about 1 hr is at risk of being legally intoxicated.

When ethanol is consumed as a single dose or as several doses over a short period of time (Fig. 3-2), the BAC curve shows a rising phase, a plateau, and a falling phase. The curve shown in Fig. 3-2 is typical for a 70-kg adult male who consumes 8 oz of 100-proof bourbon (120 ml of ethanol) over a short time. During the rising phase, ethanol absorption exceeds the rate of metabolism, primarily in the liver. About 25% of the ingested alcohol is absorbed in the stomach, and 75% is absorbed in the small intestines. The plateau occurs when the absorption rate equals the metabolic oxidation rate. During the falling phase, absorption is complete, and one can estimate the rate of ethanol metabolism in vivo from the nearly linear descending slope of the BAC curve. On average, the BAC falls by approximately 15 mg/dl per hour. In this example, the individual is

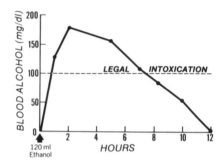

Figure 3-2. BAC versus time of consumption. (From Smith, 1981. Reprinted with permission.)

legally intoxicated less than 1 hr after consuming the ethanol and remains legally intoxicated for the next 7 hr.

Ethanol is eliminated from the body primarily by oxidation in the liver. Only about 2–10% is eliminated by all other mechanisms, including excretion in the urine, bile, sweat, and saliva and loss in expired air. The Breathalyzer test assumes that alcohol in expired air is directly proportional to the arterial BAC. The amount of alcohol oxidized per unit time is largely independent of the BAC, and this imposes an upper limit on the rate of consumption that one can sustain without succumbing to the CNS effects of alcohol metabolism.

As stated above, the average adult requires about 1–1.5 hr to metabolize 15 ml of ethanol (one 12-oz bottle of beer or 1 oz of 100-proof bourbon). Figure 3-3 shows the BAC profiles for men who consumed 1 and 2 oz of 100-proof ethanol for each 150 lb of body weight. Men who consumed ethanol at the rate of 1 oz/hr could drink for 7 hr without reaching the legal point of intoxication. When the dose was doubled to 2 oz/hr the men were legally intoxicated in about 2.5 hr.

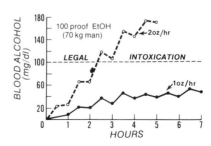

Figure 3-3. BAC versus rate of consumption. (From Smith, 1981. Reprinted with permission.)

3.3. EFFECTS OF ALCOHOL INTAKE ON FATTY ACID METABOLISM

Ethanol intake markedly inhibits fatty acid oxidation in liver (Fig. 3-4). These results were obtained from isolated rat livers perfused with ^{14}C-labeled triacylglycerols in chylomicrons. The production of ^{14}C-CO_2 measures fatty acid oxidation in mitochondria. The inhibitory effect of alcohol is clearly evident. The mechanism for this inhibition results in large part from the high NADH/NAD$^+$ ratio, which inhibits both β oxidation of fatty acids in the mitochondria and the tricarboxylic acid cycle. β Oxidation of fatty acids is inhibited because of the low level of NAD$^+$. The tricarboxylic acid cycle is inhibited as a result of (1) the inhibitory action of NADH on isocitrate dehydrogenase and α-ketoglutarate dehydrogenase and (2) a depletion of oxalacetate. The high NADH/NAD$^+$ ratio allows malate dehydrogenase to reduce oxalacetate to malate.

The inhibition of fatty acid oxidation leads to an accumulation of fatty acids, which are shunted into triacylglycerols. This increases the level of triacylglycerols in the liver and induces the liver to produce more VLDL. When the rate of synthesis of triacylglycerols exceeds the capacity of the liver to package them into VLDL, then triacylglycerol droplets accumulate in the liver and produce a fatty liver (Fig. 3-5).

The fatty acids that contribute to triacylglycerol accumulation in the liver are derived in part from mobilization, from dietary fat, and by de novo synthesis in the liver from acetyl-CoA. The major source is dietary fat. Indeed, the intake of excessive alcohol coupled with a high-fat diet aggravates the development of a fatty liver. Fatty liver is believed to enhance the development of alcoholic hepatitis, which in turn is believed to increase the chance of developing liver cirrhosis. Other consequences of the increased NADH/NAD$^+$ ratio include ketonemia, ketoacidosis, lactic acidemia, lactacidosis, hyperuricemia, and alcoholic hypoglycemia.

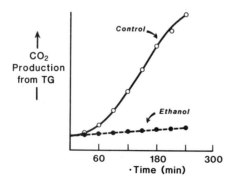

↑
CO$_2$
Production
from TG
|

Control

Ethanol

60   180   300
·Time (min)

Figure 3-4. Ethanol inhibition of fatty acid oxidation. (From Bonkowsky, 1981. Reprinted with permission.)

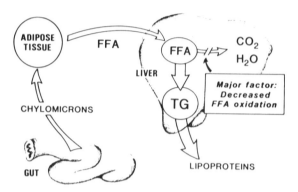

Figure 3-5. Alcohol-induced fatty liver. (From Bonkowsky, 1981. Reprinted with permission.)

Some acetyl-CoA derived from alcohol and fatty acid oxidation is shunted into ketogenesis by conversion to HMG-CoA, which is cleaved to form free acetoacetic acid. The acetoacetic acid is in large part reduced to β-hydroxybutyric acid. Excessive production of acetoacetic acid and hydroxybutyric acid (which are called ketone bodies) leads to ketosis and ketoacidosis. One consequence is hyperuricemia, which results from an inhibition of uric acid secretion in the kidney by high levels of lactic acid.

3.4. ALCOHOLIC HYPOGLYCEMIA

The high $NADH/NAD^+$ ratio leads to reduction of pyruvate to lactate, reduction of oxalacetate to malate, and reduction of DHAP to α-glycerolphosphate, thereby depriving the liver of three important intermediates for gluconeogenesis. The inhibition of gluconeogenesis and the fall in blood glucose cause hypoglycemia (Fig. 3-6). The hypoglycemia is more severe in persons who are on poor (especially carbohydrate) diets since glycogen reserves are low or depleted and cannot provide sufficient blood glucose. In alcoholics, who are usually on very poor diets, the hypoglycemia can be severe and life threatening.

3.5. ALCOHOLIC HEPATITIS AND LIVER CIRRHOSIS

Liver damage is related to the degree and duration of ethanol ingestion, although other environmental and genetic factors may modulate the development of fatty liver, hepatitis, and cirrhosis. Heavy drinking (more than 5 oz of alcohol per day) has been shown to predispose one to liver damage. The risk of liver

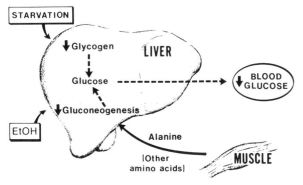

Figure 3-6. Alcoholic hypoglycemia. Dashed line indicates inhibition. (From Smith, 1981. Reprinted with permission.)

damage triples upon increasing the consumption of alcohol from 21 g/day to 40 g/day, and the risk increases 600-fold with a daily intake of 140 g of alcohol per day. Alcoholic hepatitis is a disease that takes several years to develop. It is characterized by smoldering necrosis and inflammation. The clinical symptoms are variable; persons more severely affected are typically weak, poorly nourished, febrile, and suffering abdominal pain.

Alcoholic liver injury typically occurs first and foremost in zone III of the hepatic acinus, that is, around the terminal hepatic venules or "central" veins. Fat accumulation and the histological changes of alcoholic hepatitis first appear in this region of the liver. These pathological changes have been termed central hyaline sclerosis. If progressive, the sclerotic changes compromise the lumen or the terminal hepatic venule and obstruct hepatic venous flow. There also may occur progressive collapse and fibrosis of liver parenchymal cells, with eventual drawing together of sclerotic central vein remnants and fibrotic tracts and the accumulation of collagen and connective tissue. Regenerative activity, forming nodules of hepatocytes with disordered architecture, completes the picture of cirrhosis.

Much of what is known about the sequential anatomical and biochemical changes in the pathogenesis of alcoholic liver disease and cirrhosis has come from studies on alcohol-fed baboons. The accumulation of intrahepatic fat is well documented. Three types of cells in liver can accumulate fat: lipocytes, whose normal function is fat storage; transition cells (possibly derived from lipocytes); and hepatocytes. In fatty liver, the majority of fat accumulates in hepatocytes rather than lipocytes. However, transition cells are increased in alcoholic liver injury in alcohol-fed baboons, and cell processes emanating from both transition cells and lipocytes develop microfilaments, dense bodies, and pinocytic vesicles. Also, the total area of the rough endoplasmic reticulum in transition cells is 40% greater than that of lipocytes which suggests greater protein and sterol synthetic activity. Indeed, there is a significant correlation between the percentage of transition cells present per liver and the total area of rough endoplasmic reticulum in transition cells with the overall degree of hepatic fibrosis in alcohol-fed baboons. These findings may explain the origin of perisinusoidal fibrosis in cirrhotic liver (Nutr. Rev., 1985).

The complications of cirrhosis are attributable mainly to diminished liver cell function and marked abnormalities in portal blood flow. The obstruction to the flow of portal blood through the liver leads to portal hypertension; this condition may lead to development of varices and hemorrhoids as a result of enlargement of alternative collateral channels between the splanchnic and systemic circulations. These dilated collaterals may rupture and produce gastrointestinal bleeding. The bleeding may be difficult to stop because of the accompanying defects in blood coagulation since the liver cannot produce sufficient amounts of the necessary proteins involved in clotting. Excessive bleeding of esophageal varices can be lethal.

Other complications of cirrhosis includes ascites, caused by the low osmotic pressure in the plasma as a consequence of low plasma albumin. The liver is not able to produce sufficient albumin, which maintains the osmotic pressure of the blood. Consequently, water flows from the blood into the abdomen. Renal tubular acidosis, renal failure, increased incidence of peptic ulcer, increased incidence of gallstones, diabeteslike glucose intolerance, impaired immune function, impaired leukocyte function, hypersplenism, ammonia toxicity leading to hepatic coma, and development of hepatoma are other possible complications of cirrhosis.

3.6. EFFECTS OF ALCOHOL ON THE CNS

Most of the acute pharmacological effects of ethanol are manifested through the CNS. Ethanol is a primary, functional, and irregularly descending CNS depressant, affecting the polysynaptic structures of the reticular activating system of the cortex. As the cortex is released from its integrating control, the drinker experiences the "high" or euphoric stage of intoxication. Depression of the cerebellum leads to a loss of motor coordination, and depression of the midbrain interferes with spinal reflexes and temperature regulation. These effects lead to the familiar progression from ataxia to stupor to coma, and the heavy drinker falls into a state of general anesthesia. Ultimately the medulla succumbs, and death results from a paralysis of the medullary centers and subsequent respiratory failure.

Ethanol, however, is not a useful general anesthetic agent, since its anesthetic threshold lies dangerously close to the threshold for central respiratory depression. The anesthetic property of ethanol is related to its lipid solubility and its insertion in cell membranes, where it is believed to alter the fluidity of the membranes. Some experimental evidence suggests that ethanol increases the fluidity of the neuronal cell membrane. This evidence has led to a tentative hypothesis that the increased membrane fluidity is responsible for the acute effects of alcohol.

However, long-term intake of alcohol can lead to adaptive changes in the membrane in an attempt to restore the normal fluidity. It has been speculated but not proven that these adaptive changes are in part responsible for the phenomenon of tolerance of and physical dependence on alcohol. It is thought that these changes may be produced by cells synthesizing and inserting more saturated fatty acids or inserting more cholesterol in the membrane, or a combination of these actions.

The CNS effects of excessive alcohol intake are believed to be due in part to the production of acetaldehyde, which is a very reactive molecule, reacting with amino groups of proteins, phospholipids such as phosphatidylethanolamine and phosphatidylserine, and catecholamines. Some experiments in animals have indicated that acetaldehyde condenses with catecholamines and forms isoquinoline compounds, which may be converted to morphinelike compounds (Davis and Walsh, 1970). The latter compounds may account in part for the withdrawal symptoms of alcohol intoxication. Alcohol intake may also decrease the level of cGMP in the brain. cGMP is required for certain neuroexcitatory responses in the cerebellum, mediated by acetylcholine receptors. New work by Lovinger et al. (1989) has shown that ethanol (5–50 mM range) inhibits the glutamate receptor agonist NMDA (N-methyl-D-aspartate)-activated ion current in hippocampal neurons in a linear dose-dependent manner. This inhibition of NMDA receptor activation is believed to contribute to the neural and cognitive impairments associated with intoxication. The mechanism by which alcohol inhibits the NMDA-activated ion current is not known but may be due in part to the fluidization of the neuronal membrane by alcohol, which in turn influences the glutamate receptor. The question of whether neurochemical actions are responsible for alcohol intoxication in vivo has recently been addressed by the use of animal populations displaying genetic differences in sensitivity to alcohol and benzodiazepine intoxication (Harris and Allan, 1989). These workers used inbred strains, selected lines, recombinant inbred strains, and heterogeneous stocks of mice and rats to study the effect of alcohol on γ-aminobutyric acid (GABA)-stimulated chloride channels and voltage-dependent calcium and sodium channels of isolated brain membranes. They found that genetic differences in ion channel function provide strong evidence for a role of the GABA-stimulated chloride channel in ethanol and benzodiazepine intoxication but no clear role of the calcium and sodium channel.

Alcoholics have a higher plasma level of acetaldehyde than do nonalcoholics. Blood acetaldehyde concentrations ranged from 0.11 to 0.15 mg/dl in 15 alcoholic men and from 0.04 to 0.08 mg/dl in nonalcoholics, when the BACs ranged from 1 to 400 mg/dl, after the consumption of bourbon and grain ethanol, respectively (Majchrowicz and Mendelson, 1970).

3.7. EFFECTS OF ACETALDEHYDE

Acetaldehyde is believed to be a toxic intermediate in the metabolism of alcohol. Three ranges of acetaldehyde levels in blood can be defined: the normal range, the acute aldehyde syndrome (very high levels of acetaldehyde), and chronic aldehydism (slightly elevated levels of acetaldehyde). In alcoholics, slightly elevated levels of blood acetaldehyde are found. In Orientals lacking the low-K_m ALDH, acetaldehyde accumulates and leads to the acute aldehyde syndrome (von Wartburg and Buchler, 1984).

3.8. EFFECTS OF ALCOHOL ON THE HEART

Alcohol affects many cellular substructures in the heart. In the plasma membrane, the Na^+,K^+-dependent ATPase is inhibited. Mitochondrial respiration and calcium uptake are also inhibited. Acetaldehyde has a direct effect on myocardial protein synthesis. Malfunction of the sarcoplasmic reticulum is revealed by disturbances in calcium uptake. Fatty acid oxidation is inhibited, whereas fatty acid esterification to form triacylglycerols is enhanced, mimicking what occurs in the liver (Bing, 1982). Many of these effects are believed to result from the increased fluidity of cell membranes, with alteration of the activity of enzymes and proteins in the membrane (Rubin, 1982; Katz, 1982).

Acetaldehyde is a direct vasodilator in the heart. It has a stronger action in stimulating the release of norepinephrine at adrenergic nerve terminals, which leads to vasoconstriction and increased blood pressure. This in turn leads to hypertension and bradycardia.

Chronic alcohol intake does not alter endogenous norepinephrine levels in heart but does increase turnover of norepinephrine. Chronic alcohol intake lowers the number of β-adrenergic receptors and decreases the response to isoproterenol (Pohorecky, 1982).

3.9. ALCOHOL–DRUG INTERACTIONS

The increased gastric blood flow resulting from alcohol intake, coupled with the solvency of alcohol, can produce increased rates of drug absorption and heightened drug effects. When taken as a single large dose, alcohol inhibits the drug- detoxifying enzymes in the liver and makes a person more sensitive to (less tolerant of) the action of the drug. This effect is mediated by an inhibition of the mixed-function oxidase of the endoplasmic reticulum. However, chronic intake of alcohol leads to hyperplasia of the endoplasmic reticulum and hence increased levels of the mixed-function oxidase, which allows the liver to detoxify the drug more rapidly and makes a person more resistant (more tolerant) to the action of some drugs. This effect pertains only to those drugs that are metabolized by enzymes of the endoplasmic reticulum and whose levels increase when the endoplasmic reticulum undergoes hypertrophy.

3.10. ALCOHOL WITHDRAWAL SYMPTOMS

With chronic intake of alcohol, physical dependence develops to a point where unambiguous withdrawal reactions are recognized. Minor withdrawal symptoms are characterized by insomnia, irritability, and tremor, whereas major symptoms include anxiety, agitation, sweating, delirium, and disorientation. Withdrawal seizures may occur in the 12- to 48-hr period after the start of abstinence. The combination of delirium with tremor gave rise to the name delirium tremens (DTs), which in its most serious form includes vivid, terrifying hallucinations, tachycardia, fever, sweating, and a profound hypermetabolic state. Although alcohol is a specific antagonist to the withdrawal reaction, benzodiazepines are currently the drugs of choice for preventing or aborting the abstinence syndrome. The intact alcohol molecule appears to be the addicting agent, since physical dependence also develops with tertiary butanol, which does not undergo extensive biotransformation in vivo (Bonkowsky, 1981).

3.11. ALCOHOL INTAKE AND HEART DISEASE

Moderate intake of alcohol (one or two glasses of wine per day) increases the synthesis of apoA in liver and increases the level of HDL in plasma. The HDL subfraction that is increased is called HDL-C because of its unusual cholesterol content. Whether this offers protection from or alternatively increases the risk of coronary heart disease is not known. Even this modest amount of alcohol intake increases the synthesis of VLDL in liver and hence increases the level of VLDL in plasma an effect that may be deleterious to persons with type III or IV hyperlipidemia. Therefore, this potential beneficial effect of alcohol intake varies, depending on each person's genetic make-up and ability to synthesize and degrade plasma lipoproteins. The effect of ethanol intake on plasma lipoproteins is covered in Chapter 7.

REFERENCES

Abel, E. L., 1985, Prenatal effects of alcohol on growth: a brief overview, *Fed. Proc.*, 44:2318.

Alcohol intake, blood lipids and mortality from coronary heart disease, 1984, *Clin. Nutr.*, 3:139.

Bing, R.J., 1982, Effect of alcohol on the heart and cardiac metabolism, *Fed. Proc.*, 41:2443.

Bonkowsky, H.L., 1981, *Alcohol and the Liver, Unit 2, Undergraduate Teaching Project,* Project Cork Institute, Dartmouth College, Milner-Fenwick Inc., (distributor) Timonium, Maryland.

Camargo, C.A., Jr., Williams, P. T., Vranizan, K. M., Albers, J. J., and Wood, P. D., 1985, The effect of moderate alcohol intake on serum apolipoproteins A-I and A-II. A controlled study, *J. Amer. Med. Assoc.*, 253:2854.

Changes in plasma lipoproteins due to alcohol consumption, 1985, *Nutr. Rev.*, 43:74.

Cloninger, C. R., 1987, Neurogenetic adaptive mechanisms in alcoholism, *Science*, 236:410.

Crouse, J. R., and Grundy, S. M., 1984, Effects of alcohol on plasma lipoproteins and cholesterol and triglyceride metabolism in man, *J. Lipid Res.*, 25:486.

Curran, M., and Seeman, P., 1977, Alcohol tolerance in a cholinergic nerve terminal: relation to the membrane expansion-fluidization theory of ethanol action, *Science*, 197:910.

Davis, V. E., and Walsh, M. J., 1970, Alcohol, amines, and alkaloids: a possible biochemical basis for alcohol addiction, *Science*, 167:1005.

Deitrich, R. A., and Erwin, V. G., 1984, Involvement of biogenic amine metabolism in ethanol addiction, *Fed. Proc.*, 34:1962.

Halley, S. B., and Dzvonik, M. L., 1984, Alcohol intake, blood lipids and mortality from coronary heart disease, *Clin. Nutr.*, 3:143.

Harris, R. A., and Allan, A. M., 1989, Alcohol intoxication: ion channels and genetics, *FASEB J.*, 3:1689.

Hepatic fat and alcoholic cirrhosis, 1985, *Nutr. Rev.*, 43:124.

Israel, Y., Orrego, H., Colman, J. C., and Britton, R. S., 1981, Alcohol-induced hepatomegaly: pathogenesis and role in the production of portal hypertension, *Fed. Proc.*, 41:2472.

Katz, A. M., 1982, Effects of ethanol on ion transport in muscle membranes, *Fed. Proc.*, 41:2456.

Lieber, C. S. (ed.), 1977, *Metabolic Aspects of Alcoholism,* University Park Press, Baltimore.

Lovinger, D. M., White, G., and Weight, F. F., 1989, Ethanol inhibits NMDA-activated ion current in hippocampal neurons, *Science*, 243:1721.

Majchrowicz, E., 1975, Effect of peripheral ethanol metabolism on the central nervous system, *Fed. Proc.*, 34:1948.

Majchrowicz, E., and Mendelson, J. H., 1970. Blood concentrations of acetaldehyde and ethanol in chronic alcoholics, *Science*, 168:1100.

Majchrowicz, E. M., and Noble, E. P. (eds.), 1979, *Biochemistry and Pharmacology of Ethanol,* Vols. 1 and 2, Plenum Press, New York.

Mezey, E., 1985, Metabolic effects of alcohol, *Fed. Proc.*, 44:134.

Moderate alcohol consumption increases plasma high density lipoprotein cholesterol, 1987. *Nutr. Rev.*, 45:8.

Patek, A. J., 1979, Alcohol, malnutrition, and alcoholic cirrhosis, *Am. J. Clin. Nutr.*, 32:1304.

Pohorecky, L. A., 1982, Influence of alcohol on peripheral neurotransmitter function, *Fed. Proc.*, 41:2452.

Rubin, E., 1982, Alcohol and the heart: theoretical considerations, *Fed. Proc.*, 41:2460.

Rubin, E., and Rottenberg, H., 1982, Ethanol-induced injury and adaptation in biological membranes, *Fed. Proc.*, 41:2465.

Smith, R. P., 1981, *The Biochemistry, Pharmacology, and Toxicology of Alcohols, Undergraduate Teaching Project,* Project Cork Institute, Dartmouth College, Milner-Fenwick, Inc., (distributor), Timonium, Maryland.

von Wartburg, J. P., and Buchler, R., 1984, Biology of disease. Alcoholism and aldehydism: new biochemical concepts. *Lab. Invest.*, 50:5.

Chapter 4

DISORDERS OF CHOLESTEROL METABOLISM
Cholesterol Storage Diseases

4.1. OVERALL CHOLESTEROL BALANCE IN HUMANS

Cholesterol is a white waxy sterol lipid that occurs in all animal cells. It has an important structural role in cell membranes, is the precursor for steroid hormones in the adrenal gland, and is the precursor for bile acids in the liver. Because of its insolubility in water, it is solubilized and transported in the blood as a lipoprotein complex. Cholesterol input to the body comes from the diet and endogenous synthesis, primarily by the liver. Cholesterol output occurs via secretion in the bile, conversion to bile acids, and loss from sloughing off of cells from the skin and intestines; a very small amount is lost in the urine. Lactating females also lose some cholesterol during breast feeding. The daily balance of cholesterol metabolism in humans is shown in Fig. 4-1.

An 70-kg adult human contains about 140 g of total cholesterol, of which about 8 g is present in plasma. The average daily diet contains about 500 mg of cholesterol and the liver synthesizes about 1000 mg/day. Since the daily metabolic need for cholesterol is approximately 350 mg, the balance has to be excreted, mainly via the bile.

4.2. SYNTHESIS OF CHOLESTEROL

The de novo synthesis of cholesterol begins with acetyl-CoA and occurs primarily in the liver. It is estimated that the synthesis requires about 26 separate reactions. The overall equation for cholesterol synthesis is shown in Fig. 4-2. This overall reaction scheme is provisional, since not all of the steps involved are understood. Synthesis of cholesterol involves reductive condensation, cycliza-

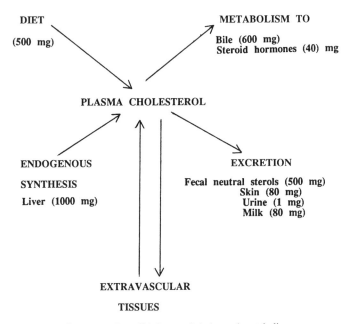

DIET

(500 mg)

METABOLISM TO

Bile (600 mg)
Steroid hormones (40) mg

PLASMA CHOLESTEROL

ENDOGENOUS

SYNTHESIS
Liver (1000 mg)

EXCRETION

Fecal neutral sterols (500 mg)
Skin (80 mg)
Urine (1 mg)
Milk (80 mg)

EXTRAVASCULAR

TISSUES

Figure 4-1. Overall balance of cholesterol metabolism.

tion, hydroxylation, shifts of H and CH_3 groups, conversion of three CH_3 groups to CO_2, and introduction and migration of double bonds. The synthesis begins in the cytosol and is completed in the endoplasmic reticulum. Some of the steps in the synthesis of cholesterol are depicted in Fig. 4-3.

The water-soluble products are made in the cytosol, whereas the lipid-soluble products are made in the endoplasmic reticulum membrane. Phosphorylated intermediates play an important role in the early part of synthesis up to the formation of farnesyl pyrophosphate. As the intermediates in the synthesis become more hydrophobic the process occurs in the endoplasmic reticulum.

The rate-limiting regulatory step in the synthesis of cholesterol is the conversion of HMG-CoA to mevalonic acid (Fig. 4-4). The enzyme catalyzing this

$$18\text{Acetyl-CoA} + 15\text{NADPH} + 6\text{NADH} + 3\text{NAD}^+ + 4O_2 + 21H^+ \text{---------}>$$

$$\text{cholesterol} + 9CO_2 + 15\text{NADP}^+ + 6\text{NAD}^+ + 3\text{NADH} + 6\text{PPi} + 3H^+$$

Figure 4-2. Overall synthesis of cholesterol from acetyl-CoA. This is a provisional equation, since not all of the steps in the synthesis are known. The equation is not balanced.

2Acetyl-CoA -------> acetoacetyl-CoA

Acetyl-CoA + acetoacetyl-CoA --------> HMG-CoA (C-6)

HMG-CoA + 2NADPH -------> mevalonic acid

Mevalonic acid + ATP --------> mevalonate phosphate

Mevalonate phosphate + ATP --------> mevalonate pyrophosphate

Mevalonate pyrophosphate ----> isopentenyl pyrophosphate + CO_2

Isopentenyl pyrophosphate <-----> dimethylallyl pyrophosphate

   C-5)   (C-5)

2C-5 units ---------> geranyl pyrophosphate (C-10)

Geranyl pyrophosphate + C-5 -----> farnesyl pyrophosphate

   (C-10)   (C-15)

2Farnesyl pyrophosphate + NADPH ---------> squalene (C-30)

Squalene + NADPH + O_2 ---------> lanosterol (C-30)

Lanosterol + $3O_2$ + 2NADPH + $NADP^+$ + 2NADH ---> ---> --->

                  ---> ---> ---> cholesterol (C-27) + $3CO_2$

Figure 4-3. Biosynthesis of cholesterol.

reaction is HMG-CoA reductase (HMGR), the synthesis of which is stimulated by insulin. The short-term and long-term regulation of HMGR is very similar to that of acetyl-CoA carboxylase, which is the rate-limiting enzyme in the synthesis of fatty acids. Insulin stimulates the dephosphorylation of HMGR, whereas hormones such as epinephrine and glucagon stimulate the phosphorylation of HMGR. Phosphorylation inhibits and dephosphorylation activates the enzyme. Insulin appears to activate the phosphatase that dephosphorylates HMGR and also inhibits the action of a protein that is an inhibitor of the phosphatase. The biochemical mechanisms of action of insulin remain to be determined.

Mevalonic acid is phosphorylated by two successive reactions to isopentenyl pyrophosphate. Two separate enzymes are involved, and ATP provides the phosphate groups. Isopentenyl pyrophosphate and its isomer dimethylallyl pyro-

                              HMGR
HMG-CoA + 2NADPH -----------> mevalonic acid + $2NADP^+$ 2H+

Figure 4-4. Conversion of HMG-CoA to mevalonic acid. This is the rate-limiting step in the synthesis of cholesterol catalyzed by HMGR.

phosphate are the C-5 condensing units used for the synthesis of the C-10 and C-15 intermediates, called geranyl pyrophosphate and farnesyl pyrophosphate, respectively. Two farnesyl pyrophosphate units condense to form presqualene pyrophosphate in a reaction involving the loss of one H atom from the C-1 of one of the two molecules of farnesyl pyrophosphate. Presqualene pyrophosphate undergoes rearrangement and subsequent reduction with NADPH to the C-30 terpene squalene. Squalene is converted to squalene-2,3-oxide by squalene epoxide, which utilizes molecular oxygen and NADPH. Squalene oxide is then cyclized to lanosterol by squalene oxide cyclase. This cyclization reaction is postulated to be initiated by attack of a proton on the oxide ring and is followed by concerted electron shifts leading to ring closure. A transient carbonium at C-20 is believed to be an intermediate. Lanosterol is derived from squalene oxide by a series of concerted hydride and methyl shifts and elimination of a proton from C-9. Through a series of complex steps, lanosterol (C-30) is ultimately converted to cholesterol (C-27).

Cholesterol and triacylglycerols are very insoluble in water. For this reason, their synthesis occurs with enzymes localized on cell membranes and is synchronized with the synthesis of phospholipids and specific proteins; these lipids are precisely assembled into VLDL and HDL in the liver. The lipoproteins are then secreted into the bloodstream where they carry cholesterol and phospholipids to other cells of the body.

However, not all of the cholesterol is used for the synthesis of lipoproteins; a substantial part is used for the synthesis of bile acids in the liver. Since cholesterol cannot be oxidized to carbon dioxide and water as are the fatty acids, the liver must have a mechanism for disposing of excess cholesterol. Conversion to bile acids serves this function very well, especially since the bile acids are not just waste products but rather are stored in the gall bladder and used for the emulsification of dietary lipids (Chapter 1). The liver also secretes free cholesterol and phosphatidylcholine in bile. Indeed, the molar compositions of cholesterol, bile acids, and phosphatidylcholine are critical in keeping cholesterol from crystallizing and forming gallstones.

4.3. CONVERSION OF CHOLESTEROL TO BILE ACIDS

In Chapter 1 the bile acids were discussed with respect to their role in lipid digestion. The conversion of cholesterol to bile acids is shown in Fig. 4-5. The rate-limiting step in the conversion of cholesterol to bile acids is the hydroxylation at C-7, which is catalyzed by 7-α-cholesterol hydroxylase. The reaction requires NADPH and O_2. Vitamin C is believed to stimulate the activity of this enzyme. The enzyme is regulated is by negative feedback control by the bile acids. Regulation is also effected by phosphorylation-dephosphorylation as in the

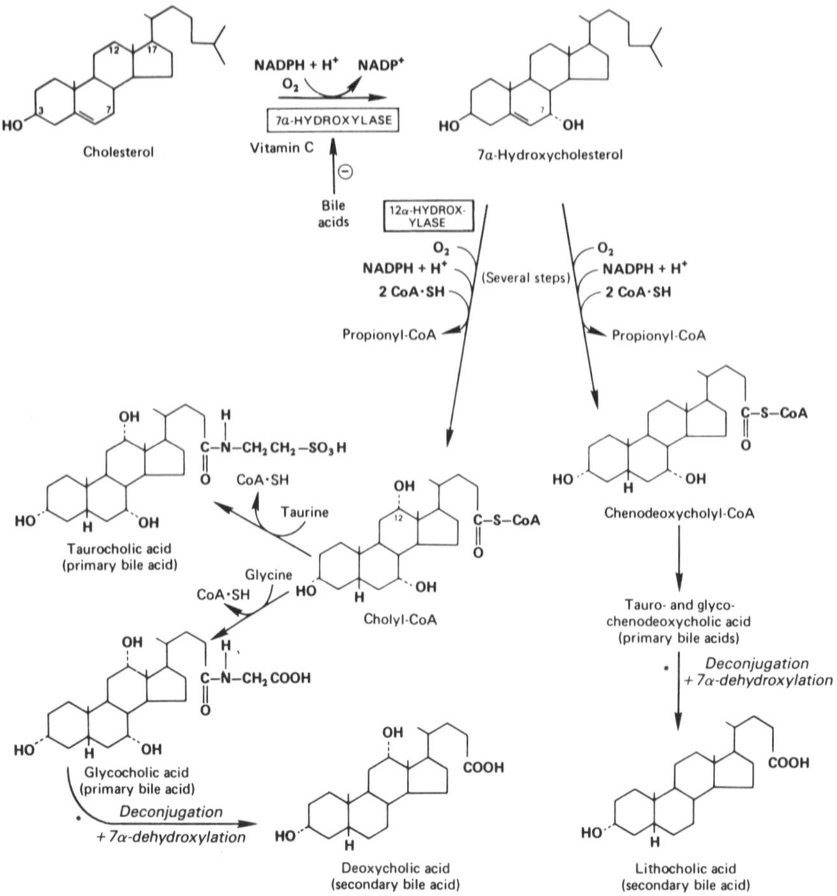

Figure 4-5. Conversion of cholesterol to bile acids. (From Murray et al., 1988. Reproduced with permission.)

case of acetyl-CoA carboxylase and HMGR except that the phosphorylated form rather than the dephosphorylated form of the enzyme is active. Peroxisomes are believed to play an important role in bile acid synthesis. These cell particles contain both NADPH cytochrome P-450 reductase and NADH cytochrome P-450 reductase. Peroxisomes are able to oxidize 3-α, 7-α, 12-α-trihydroxy-5-β-cholestanoic acid to cholic acid. Hydroxylation of cholesterol is necessary for its conversion to bile acids. Hydroxylation occurs both in the endoplasmic reticulum and in the mitochondria, utilizing a microsomal NADPH cytochrome P-450 reductase or a mitochondrial NADPH ferrodoxin reductase.

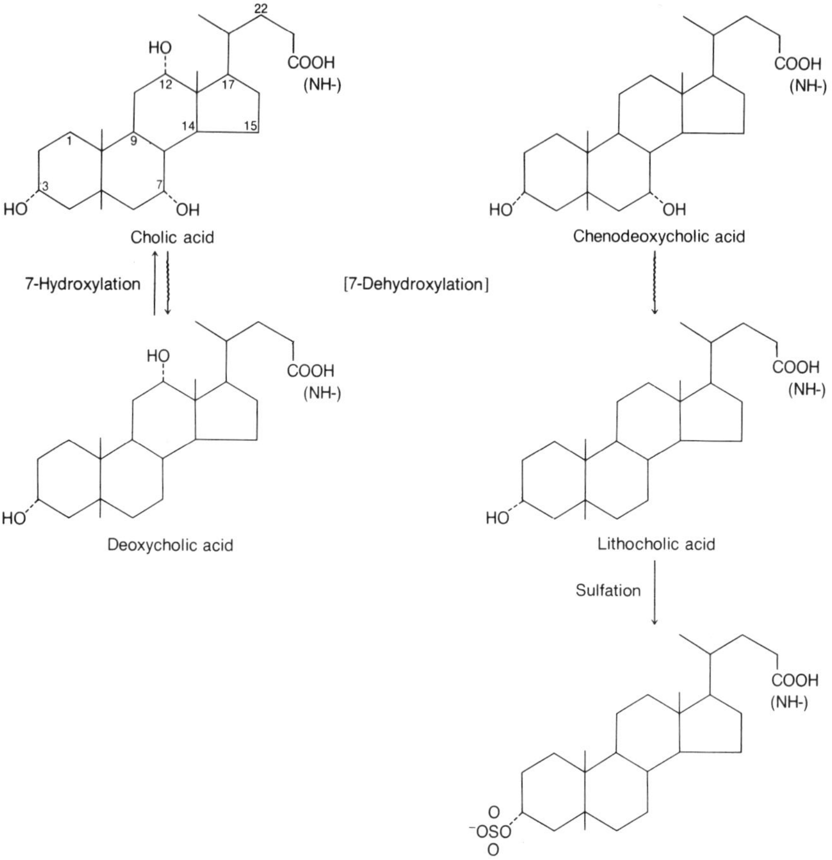

Figure 4-6. Metabolism of bile acids. (NH⁻) signifies the glycine and taurine groups of the conjugated bile acids. (From Devlin, 1982. Reproduced with permission.)

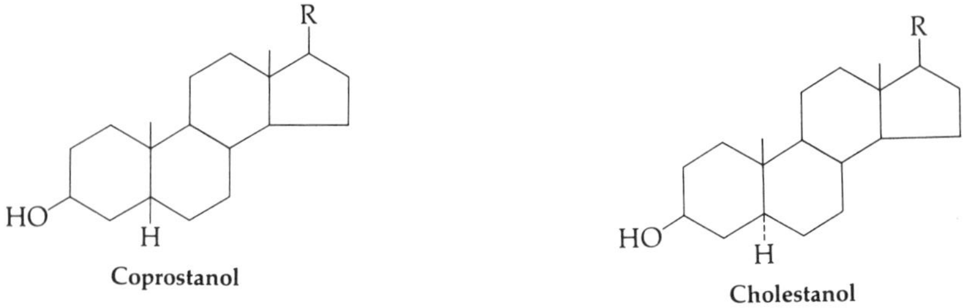

Figure 4-7. Structures of cholestanol and coprostanol.

The primary bile acids, which are made in the liver, are cholic and chenodeoxycholic acid. These acids react with two small polar molecules, glycine and taurine, to yield the conjugated bile acids. Taurine and glycine are linked via an amide bond to the carboxyl group of the bile acids. The addition of taurine and glycine makes the conjugated bile acids more polar, stronger acids and better emulsifying agents.

When the gallbladder contracts under the influence of the hormone CCK-pancreozymin, bile is ejected into the small intestines, where the bile acids emulsify the ingested dietary lipids. As the bile acids pass along the intestinal lumen, they are subject to modification by intestinal bacteria. These bacteria deconjugate and dehydroxylate the primary bile acids and their conjugates to form the secondary bile acids, deoxycholic acid and lithocholic acid (Fig. 4-6). Sulfation of lithocholic acid also occurs in the small intestine. As mentioned previously, about 90% of the bile acids are reabsorbed via the enterohepatic system and the rest are lost in the feces.

Intestinal bacteria also convert an appreciable amount of cholesterol (dietary or from bile) into two other sterols, cholestanol and coprostanol (Fig. 4-7). These sterols are poorly absorbed and are lost in the feces.

In the adrenal gland, cholesterol is stored as cholesterol ester, which serves as the substrate for the synthesis of adrenal steroids. In the skin, under the influence of ultraviolet light, some cholesterol is converted to vitamin D. Although these syntheses are very important physiologically, they represent only a very small amount of the mass of the total cholesterol that is metabolized in the body.

4.4. ABNORMALITIES OF CHOLESTEROL METABOLISM

Hypercholesterolemia denotes abnormally high levels of cholesterol in blood plasma. Elevated levels of plasma cholesterol are due primarily to increased levels of LDL or VLDL. Elevated levels of these cholesterol-containing lipoproteins are associated with atherosclerosis (discussed in Chapter 6).

4.4.1. Wolman's Disease

Wolman's disease is a rare familial disease characterized by the accumulation of large amounts of cholesterol esters and triacylglycerols in liver, spleen, adrenal glands, hematopoietic system, and small intestine. Gastrointestinal symptoms, hepatosplenomegaly, steatorrhea, and adrenal calcification usually occur in the first weeks of life, and death occurs by the age of 6 months in homozygous patients. Heterozygous patients have no apparent clinical symptoms, although the activity of the lipase is decreased by 50%. Patients have a marked deficiency of acid lipases, which catalyze the hydrolysis of cholesterol esters and triacylglycerols. Studies on isolated fibroblasts and leukocytes show that these lipids accumulate in lysosomes of the cells. These cells can take up LDL from the plasma but cannot hydrolyze cholesterol esters released from LDL to free cholesterol. Thus, the free cholesterol concentration in the cell does not rise sufficiently to inhibit HMGR or to suppress the synthesis of LDL receptors. This signals the cell to take up more LDL, and thus the level of cholesterol esters in the cell rises very markedly.

4.4.2. Cholesterol Ester Storage Disease

Cholesterol ester storage disease is a rare familial disease in which the liver is enlarged and contains high levels of cholesterol esters as a result of a deficiency of cholesterol esterase (an acid lipase). Cholesterol esters are introduced into cells mainly as LDL. The LDL particles are normally degraded in lysosomes. In the absence of cholesterol esterase, however, cholesterol esters accumulate in secondary lysosomes, eventually filling the cell to produce a "foam" cell. This process leads to cell dysfunction. Many patients may be asymptomatic with normal liver function. However, hepatosplenomegaly and elevated serum cholesterol levels are common, and death usually occurs before age 20 as a result of either liver failure or coronary heart disease. Diagnosis is made by liver biopsy, analysis of cholesterol ester content, and assay of cholesterol esterase. Patients may have a complete or partial deficiency of the acid lipase. Some patients with cholesterol ester storage disease may survive up to age 40. Older patients may develop severe atherosclerosis, portal hypertension, splenomegaly, and esophageal varices.

Enzyme replacement therapy in fibroblasts from a patient with cholesterol ester storage disease has been reported by Poznansky et al. (1989). These workers conjugated the enzyme cholesterol esterase to either insulin or apoB. Binding of these conjugates to their respective membrane receptors (i.e., the insulin receptor or the LDL receptor) allows the esterase to enter the cell via receptor-mediated endocytosis, where it becomes associated with the lysosomes and then degrades cholesterol esters. Whether this technique works in vivo remains to be determined. To be effective in vivo, the enzyme conjugate must be targeted specifically to the cells that are deficient in the enzyme and the conjugate must be protected from proteolysis before it finds the target cell. In addition, the conjugate must not elicit unfavorable immune reactions in the patient.

4.4.3. Cerebrotendinous Xanthomatosis

Cerebrotendinous xanthomatosis is a familial disease first described in 1937. It is characterized by progressive cerebellar ataxia, dementia, cataracts, and tendon xanthomas. Cholesterol and cholestanol accumulate in the nervous system and tendons. The primary defect is a deficiency of a mitochondrial steroid, 26-hyroxylase, involved in the pathway for the formation of bile acids from cholesterol. This process leads to the accumulation of 7-α-hydroxy-4-cholesten-3-one, which is converted to cholestanol. Onset of symptoms is unpredictable and usually does not occur until age 10. The disease develops during adolescence and young adulthood and is characterized by severe spasticity, ataxia, cataracts, and xanthomas (cholesterol deposits) in tendons. Mental retardation occurs in the early stage of the disease, and death usually occurs by age 40–50.

4.4.4. β-Sitosterolemia with Xanthomatosis

The rare disorder β-sitosterolemia with xanthomatosis was first described in 1974. It is associated with tendinous and tuberous xanthomas and high levels of plant sterols (especially β-sitosterol) in plasma, erythrocytes, adipose tissue, and skin. This disease appears to be caused by an increased absorption of plant sterols in the intestine. Whereas the normal level of total plant sterols in plasma is <0.9/dl, afflicted patients have levels as high as 19–37 mg/dl. The molecular defect leading to the high absorption of these plant sterols is not known. Treatment of the disease consists of reducing the dietary intake of foods rich in plant sterols.

4.4.5. Pseudohomozygous Familial Hypercholesterolemia

Persons afflicted with pseudohomozygous familial hypercholesterolemia, a rare disorder identified in 1974, have severe hypercholesterolemia (350–600 mg/dl), normal plasma triacylglycerol levels, and cutaneous planar xanthomas. The parents of these patients have normal plasma cholesterol levels. Restriction of dietary cholesterol or the use of cholestyramine is very effective in the treatment of this disorder. The facts that both parents have normal plasma cholesterol levels and diet is effective in markedly lowering the cholesterol level of afflicted persons distinguish this disease from homozygous type IIa hypercholesterolemia. Moreover, these patients have normal levels of LDL receptors, unlike those with type IIa hypercholesterolemia, who have very low levels.

4.4.6. Familial LCAT Deficiency

Identified in 1967, familial LCAT deficiency is a rare disease characterized by a marked deficiency of the enzyme LCAT. LCAT catalyzes the transfer of a fatty acid (usually polyunsaturated) from lecithin (phosphatidylcholine) to cholesterol and generates lysolecithin (lysophosphatidylcholine) and cholesterol ester (Fig. 4-8).

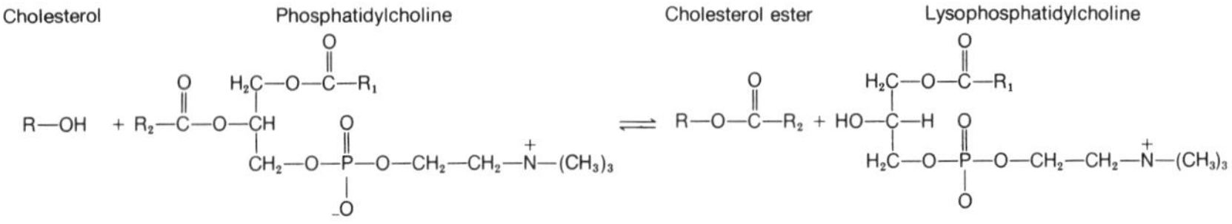

Figure 4-8. Reaction catalyzed by LCAT. (From Devlin, 1982. Reproduced with permission.)

The function of LCAT in lipoprotein metabolism is discussed in Chapter 5. LCAT allows extra cholesterol in cells to be esterified and then transported by HDL, via reverse cholesterol transport, back to the liver. Patients with LCAT deficiency have abnormal HDL particles that arrange themselves as stacks of disks or in rouleaux aggregates. Very likely, these HDL aggregates form because the HDL are low in cholesterol esters and high in free cholesterol. These abnormal HDL give rise to an opalescent plasma. All of the major plasma lipoproteins also are abnormal. The clinical features of LCAT deficiency include corneal opacities, albuminuria, and anemia, associated with abnormal target red cells. There is a very high ratio of free cholesterol to cholesterol ester in the plasma.

4.5. CHOLELITHIASIS (GALLSTONES)

It is estimated that between 16 million and 20 million Americans have gallstones. Gallstones are very common in the Pima Indians, especially the females, about 80% of whom are afflicted with cholelithiasis. Seventy-five percent of gallstones are cholesterol stones, and 25% are pigmented stones containing bile pigments and calcium. Approximately 400,000 cholecystectomies (gallbladder operations) are performed in the United States each year.

Bile is the major route for cholesterol loss from the liver. Cholesterol is maintained in solution by the formation of mixed micelles of bile acids and phospholipid, especially lecithin (phosphatidylcholine). Actually, supersaturation of cholesterol prevails in hepatic bile and to some extent in gallbladder bile. Cholesterol stones develop in the gallbladder under a variety of conditions. When bile is secreted by the liver, it is supersaturated with cholesterol. A delicate balance of the molar proportions of bile acids, lecithin, cholesterol, and water allows the cholesterol to stay in solution. An overactivity of HMGR or a depressed activity of 7-α-cholesterol hydroxylase can lead to excessive production of cholesterol and a deficiency of bile acids. This imbalance upsets the delicate balance of the lipid composition of bile and enhances the formation of cholesterol stones. The constant flow of bile from the liver to the hepatic duct helps prevent the crystallization of cholesterol in the hepatic duct. However, when bile is stored in the gallbladder, it remains stagnant for fairly long periods of time; at the same time, the gallbladder modifies the bile composition by absorbing or secreting water and lipids. In normal individuals, gallbladder bile is less supersaturated than hepatic bile, but gallstones can form as a consequence of complicated events that allow for nucleation processes leading to crystallization of cholesterol. Human gallbladder mucin binds biliary lipids and promotes formation of cholesterol crystal in model bile in vitro systems. There may also be other factors in the gallbladder that either enhance or inhibit cholesterol crystal nucleation. Groen et al. (1989) recently reported that human bile contains a concanavalin A-binding protein that stimulates cholesterol nucleation in phospholipid–cholesterol vesicles by increasing the amount of vesicular cholesterol and phospholipid and by inducing nucleation of cholesterol from these vesicles.

The lecithin/bile acid + lecithin molar ratio and the total lipid content of bile are two important factors involved in keeping cholesterol in solution but still in a supersaturated state. The importance of bile acids is seen by the success in partially dissolving cholesterol stones by giving patients 1–2 g of chenic acid (chenodeoxycholic acid) orally for several months to 1 or 2 years. Chenic acid treatment is reported to decrease the activity of HMGR and increase the activity of 7-α-cholesterol hydroxylase. Ursodeoxycholic acid given orally, also leads to unsaturation of hepatic bile with respect to cholesterol and a decrease in hepatic cholesterol. However, the unsaturation of hepatic bile does not appear to be related to a decrease in hepatic HMGR as is seen with chenic acid therapy. Ursodeoxycholic acid has been used because it appears to be more effective at lower doses than chenic acid and has less severe side effects such as diarrhea and elevated serum transaminase.

A new treatment for removing cholesterol stones in the gallbladder by ultrasound is currently under investigation in Germany and the United States. This procedure, called extracorporeal shock wave lithotripsy, involves the fragmentation of certain types of cholesterol stones by a specially devised ultrasound probe. Patients require only local analgesia for this treatment. The treatment is applicable to patients who have a functional gallbladder and no more than 3 noncalcified cholesterol stones with a diameter of less than 2.5 cm. Side effects include visceral pain and a 2% incidence of pancreatitis.

REFERENCES

Ahlberg, J., Angelin, B., Bjorkhem, I., Einarsson, K., and Leijd, B., 1979, Hepatic cholesterol metabolism in normo- and hyperlipidemic patients with cholesterol gallstones, *J. Lipid Res.,* 20:107.

Ahlberg, J., Angelin, B., and Einarsson, K., 1981, Hepatic 3-hydroxy-3-methylglutaryl coenzyme A reductase activity and biliary lipid composition in man: relation to cholesterol gallstone disease and effects of cholic acid and chenodeoxycholic acid treatment, *J. Lipid Res.,* 22:410.

Angelin, B., Ewerth, S., and Einarsson, K., 1983, Ursodeoxycholic acid treatment in cholesterol gallstone disease: effects on hepatic 3-hydroxy-3-methylglutaryl coenzyme A reductase activity, biliary lipid composition, and plasma lipid levels, *J. Lipid Res.*, 24:461.

Carey, M. C., 1978, Critical tables for calculating the cholesterol saturation in native bile, *J. Lipid Res.*, 19:945.

Carey, M. C., and Small, D. M., 1978, The physical chemistry of cholesterol solubility in bile, *J. Clin. Invest.*, 61:998.

Devlin, T. M., 1982, *Textbook of Biochemistry with Clinical Correlations*, John Wiley & Sons, New York.

Groen, A. K., Ottenhoff, R., Jansen, P. L. M., van Marle, J., and Tytgat, G. N. J., 1989, Effect of cholesterol nucleation-promoting activity on cholesterol solubilization in model bile, *J. Lipid Res.*, 30:51.

Gutierrez, C., Okita, R., and Krisans, S., 1988, Demonstration of cytochrome reductases in rat liver peroxisomes: biochemical and immunochemical analyses, *J. Lipid Res.*, 29:613.

Murray, R. K., Granner, D. K., Mayes, P. A., and Rodwell, V. W., 1988, *Harper's Biochemistry*, 21st ed., Appleton & Lange, Norwalk, Connecticut.

Poznansky, M. J., Hutchison, S. K., and Davis, P. J., 1989, Enzyme replacement therapy in fibroblasts from a patient with cholesteryl ester storage disease, *FASEB J.*, 3:152.

Reynier, M. O., Montet, J. C., Gerolami, A., Marteau, C., Crotte, C., Montet, A. M., and Mathieu, S., 1981, Comparative effects of cholic, chenodeoxycholic, and ursodeoxycholic acids on micellar solubilization and intestinal absorption of cholesterol, *J. Lipid Res.*, 22:467.

Sabine, J.R., 1977, *Cholesterol*, Marcel Dekker, New York.

Salvioli, G., Igimi, H., and Carey, M. C., 1983, Cholesterol gallstone dissolution in bile. Dissolution kinetics of crystalline cholesterol monohydrate by conjugated chenodeoxycholate-lecithin and conjugated ursodeoxycholate-lecithin mixtures: dissimilar phase equilibria and dissolution mechanisms. *J. Lipid Res.*, 24:701.

Smith, B. F., 1987, Human gallbladder mucin binds biliary lipids and promotes cholesterol crystal nucleation in model bile, *J. Lipid Res.*, 28:1088.

Smith, E. L., Hill, R. L., Lehman, I. R., Lefkowitz, R. J., Handler, P., and White, A., 1983a, *Principles of Biochemistry: Mammalian Biochemistry*, 7th ed., McGraw-Hill Book Co., New York.

Smith, E. L., Hill, R. L., Lehman, I. R., Lefkowitz, R. J., Handler, P., and White, A., 1983, *Principles of Biochemistry: General Aspects*, 7th ed., McGraw-Hill Book Co., New York.

Stryer, L., 1988, *Biochemistry*, W. H. Freeman & Co., San Francisco.

Suckling, K. E., and Stange, E. F., 1985, Role of acyl-CoA: cholesterol acyltransferase in cellular cholesterol metabolism, *J. Lipid Res.*, 26:647.

Chapter 5

DISORDERS OF LIPOPROTEIN METABOLISM
Dyslipoproteinemias

5.1. PLASMA LIPOPROTEINS AND APOPROTEINS: GENERAL ASPECTS

Plasma lipoproteins represent a variety of large heterogeneous molecular aggregates that have the major function of transporting the water-insoluble lipids (particularly triacylglycerols, cholesterol, and cholesterol esters) in the blood. Lipoproteins are dynamic molecules in a constant state of synthesis and degradation, actively exchanging certain lipids and proteins with each other.

The lipoproteins are separated on the basis of density, charge, or size. On the basis of density they are named chylomicrons, very-low-density lipoproteins (VLDL), intermediate-density lipoproteins (IDL), and high-density lipoproteins (HDL). Chylomicrons, made in the small intestines, are called exogenous lipoproteins. VLDL, made in the liver, are called endogenous lipoproteins. An electron micrograph of human plasma lipoproteins is shown Fig. 5-1. The characteristics of the lipoproteins are given in Table 5-1. (The average values obtained vary among laboratories because of different methods used to prepare the lipoproteins and because of the inherent heterogeneity of these large macromolecules.)

Chylomicrons and VLDL are high in lipid and low in protein. These lipoproteins are the major carriers of triacylglycerols (also called triglycerides) in the blood. Chylomicrons carry triacylglycerols from the intestines to the liver, whereas VLDL carry triacylglycerols (and some cholesterol) from the liver to other tissues of the body. LDL are small particles that are high in cholesterol and cholesterol esters; they are the principal carriers of these lipids in the blood. HDL are the smallest particles and contain the highest content of protein. Chylomicrons are the largest and lightest particles and give the plasma a milky appearance. VLDL and IDL, because of their intermediate size and density, give the

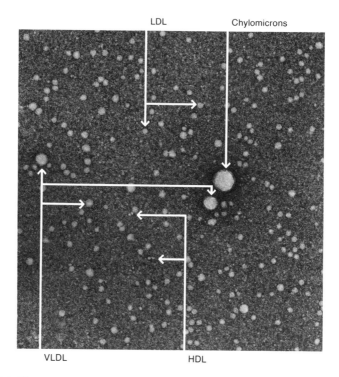

Figure 5-1. Electron micrograph of plasma lipoproteins. (From Gotto, 1980. Reproduced with permission.)

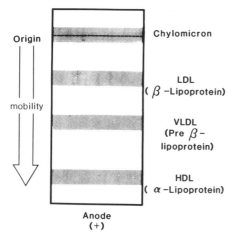

Figure 5-2. Separation of plasma lipoproteins by electrophoresis. (From Champe and Harvey, 1987. Reproduced with permission.)

Table 5-1. Characteristics of the Lipoproteins of Plasma of Humans[a]

							Composition				
								Percentage of total lipid			
Fraction	Source	Diameter (nm)	Density[b]	Sf[c]	Protein (%)	Total lipid (%)	Triacylglycerol	Phospholipid	Cholesteryl ester	Cholesterol (free)	FFA[d]
Chylomicrons	Intestine	100–1000	<0.96	>400	1–2	98–99	88	8	3	1	
VLDL	Liver and intestine	30–90	0.96–1.006	20–400	7–10	90–93	56	20	15	8	1
IDL	VLDL and chylomicrons	25–30	1.006–1.019	12–20	11	89	29	26	34	9	1
LDL		20–25	1.019–1.063	2–12	21	79	13	28	48	10	1
HDL-2	Liver and intestine VLDL?	10–20	1.063–1.125		33	67	16	43	31	10	
HDL-3	Chylomicrons?	7.5–10	1.125–1.210		57	43	13	46	29	6	6
Albumin–FFA	Adipose tissue		>1.2810		99	1	0	0	0	0	100

[a]From Murray et al., 1988. Reproduced with permission.
[b]Very-high-density liprotein is a minor fraction occurring at a density of 1.21–1.25.
[c]Sf, Sedimentation flotation constant.
[d]FFA, Free fatty acids.

Table 5-2. Some Key Characteristics of Apoproteins[a]

Apoprotein	Major lipoprotein	Mol. wt. ($\times 10^3$)	Site of synthesis	Serum concentration (mg/dl)	Function
A-I	HDL	28.3	Intestine Liver	100–150	Cofactor of LCAT; structural protein of HDL
A-II	HDL	17.0	Intestine Liver?	30–50	Phospholipid-binding properties; structural protein of HDL
A-IV	Chylo	46.0	Intestine	15	Unknown
B-100	VLDL IDL LDL	549	Liver	80–100	Intracellular formation or transcellular transport of VLDL; receptor interaction (LDL) with "apoBE receptor cells"
B-48	Chylo	265	Intestine Liver?	?	Intracellular formation or transcellular transport of chylomicrons and VLDL; receptor interaction with "apoBE receptor cells"?
C-I	HDL	6.5	Liver	<10	Cofactor of LPL of adipose tissue (particularly saturated fatty acids)
C-II	Chylo VLDL, HDL	8.8	Liver	3–8	Cofactor of LPL of adipose tissue
C-III$_{0,1,2}$	Chylo VLDL HDL	8.9	Liver	8–15	Unknown; may play a role in uptake of triglyceride-rich lipoproteins by cells
D	HDL	20	?	10	Unknown; transfer of ester cholesterol from one lipoprotein to another (?)
E	Chylo VLDL HDL	39.0	Liver	3–5	Receptor interaction with "apoBE receptor cells" and with hepatic apoE receptors; inhibitor of LPL of adipose tissue (?)
F	HDL	30.0	?	?	
G	VLDL	75.0	?	?	
H	Chylo	45.0	?	?	

[a]From Naito, 1988. Reproduced with permission.

plasma a turbid appearance. LDL and HDL, even when present in abnormally high levels, do not confer turbidity or milkiness to the plasma. High levels of LDL are associated with premature atherosclerosis and coronary heart disease, whereas high levels of HDL provide protection from these diseases.

The lipid-free proteins of the lipoproteins are called apoproteins or apolipoproteins. Some key characteristics of the apoproteins that are found in the various lipoproteins are shown in Table 5-2. Because the lipoproteins are dynamic molecules undergoing constant synthesis and degradation and interchanging some lipids and proteins with each other, the apoprotein composition of each lipoprotein is variable (with the possible exception of LDL, which contains only apoB-100). It is also possible that the protein composition can undergo some alteration during isolation, as a result of exchange processes. The apoproteins are designated by capital letters A–H. There are also subclasses of each apoprotein. These are designated apoA-I, apoA-II, apoA-III, apoA-IV, apo B-48, apoB-100, apoC-I, apoC-II, apoC-III, and apoe-1 through apoE-6. The apoproteins vary in size from relatively small (mol. wt. 7000) to large (mol. wt. 516,000) polypeptides. Some of the apoproteins have been isolated in more highly purified forms and their cDNAs isolated; their molecular weights have been revised accordingly. Thus the molecular weight of apoB-100 is now given as 514,000 and that of apoB-48 is 246,000. The molecular weights of the other apoproteins are as follows: apoA-I, 28,000; apoA-II, 17,400; apoA-IV, 46,000; apoC-I, 6600; apoC-II, 8824; apoC-III, 8750; apoD, 22,000; and apoE, 34,145.

Lipoproteins are separated and analyzed by either ultracentrifugation or electrophoresis. Ultracentrifugation separates the lipoproteins on the basis of density whereas electrophoresis separates them on the basis of electric charge. Electrophoresis can be carried out either on paper or on agarose gels. A typical electrophoretic pattern is shown in Fig. 5-2.

By ultracentrifugation, the lipoproteins are classified as chylomicrons, VLDL, IDL, LDL, and HDL. By paper electrophoresis they are grouped as chylomicrons, preβ, broad β, β, and α, respectively. Newer methods more applicable to clinical laboratories use specific columns or specific precipitating agents to separate the lipoproteins. These methods have been developed to ascertain the HDL/LDL ratio in patients, since this ratio has predictive value in assessing risk of coronary heart disease (see Chapter 6).

5.2. STRUCTURE AND FUNCTION OF APOPROTEINS

Although the structure and metabolism of the plasma lipoproteins have been the subject of intense investigation for over 50 years, it is only in the last 20 years that the primary structure of the apoproteins has been elucidated and in the last 10 years that their biosynthesis has been studied at the molecular level. In very recent years, information on the evolutionary relationships of the apoproteins has been gained through statistical analysis of DNA and protein sequencing data. These studies have uncovered unique features of internal repeats and the stringency of the requirements in the individual structural domains of the various

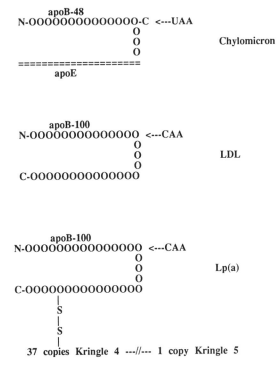

Figure 5-3. Schematic models of the apoproteins of chylomicrons, LDL, and LP(a). N and C represent the amino- and carboxy-terminal ends of the polypeptide chains. UAA and CAA designate the anticodons of the mRNA for apoB-48 and apoB-100. In apoB-48, apoE is presumed to take the place of the carboxy-terminal portion of the polypeptide of apoB-100. (Modified from Brown and Goldstein, 1987.)

apoproteins. Li et al. (1988) and Lusis (1988) have written excellent review articles on this subject.

The apoproteins appear to be synthesized as preapoproteins that undergo processing to apoproteins. Probably all of the apoproteins are either glycosylated, acylated, or phosphorylated. Some remain fixed in the lipoprotein, whereas others undergo exchange reactions between the different lipoproteins. The apoproteins vary in structure and function in different animals. They also vary widely in size (from 7000 to 500,000 molecular weight), and all appear to occur in several isoforms. ApoA, apoB, apoC, and apoE lipoproteins have been studied very extensively, but little is known about ApoD and apoF.

A basic function of all the apoproteins is to bind and transport lipids in the blood. Segrest et al. (1974) have proposed that specialized regions of the apoproteins contain amphipathic helical regions that interact with lipids. An important feature of the amphipathic helix is the presence of two clearly defined faces, one hydrophobic and the other hydrophilic. The hydrophobic face is inserted between the fatty acyl chains of the phospholipid molecules; the hydrophilic face interacts with the phospholipid polar head group and the aqueous phase. Another feature of this model is the distribution of charged amino acids on the polar face. The negatively charged residues such as glutamic and aspartic acid, tend to occur along the center of the polar face whereas the positively charged residues lysine and arginine are located on the lateral edges of the polar face.

The interaction between lipids and apoproteins differs from that between lipids and membrane-spanning proteins. In the latter case, long segments of exclusively nonpolar amino acid residues facilitate the actual insertion of the polypeptide into the hydrophobic environment. In contrast, the hydrophilic residues on one side of the amphipathic helix are in contact with the surrounding aqueous phase and keep the apoproteins at the surface of the lipoprotein particle. Such a location facilitates transfer between lipoprotein particles and interaction with enzymes (lipases and LCAT) and surface receptors.

ApoB-100 is unique among the apoproteins in having a relatively high β-sheet content. Unlike other apoproteins, apoB-100 on LDL does not transfer among the other lipoprotein particles. This observation suggests that apoB-100 binds lipid by hydrophobic β-sheet regions.

Apoproteins of the various lipoproteins have specific functions (Table 5-2). Apob-100 and apoE are recognition molecules for specific membrane receptors (the LDL receptor, also called the apoB/E receptor) located on the surface of the plasma membrane of cells that take up and utilize the lipoproteins. This receptor-mediated uptake regulates the level of these lipoproteins in plasma. Another important function of the apoproteins is their activation or inhibition of the enzymes involved in metabolism of the lipoproteins. The function of apoB-48 appears to be the enhancement of binding of apoE to the LDL receptor.

ApoA-I, apoA-II, and apoA-IV (and possibly apoC-I and apoC-III) are activators of LCAT and thus regulate the transfer of cholesterol esters between the lipoproteins. ApoC-II is an activator of lipoprotein lipase (LPL) and therefore regulates the breakdown of chylomicrons and VLDL in plasma.

The chromosomal organization of the genes for lipid metabolism and for lipid transport in humans is presented in Table 5-3. In many cases the genes are clustered. The genes for the major peripheral apoproteins (those other than apoB) are members of a family derived from a common ancestral sequence. The genes are partly dispersed on three separate chromosomes, but there is a tight cluster of genes for apoE, apoC-I, and apoC-II on chromosome 19. The chromosome 19 cluster contains two copies of the apoC-I gene, although one of these could be a pseudogene. Certain members of the gene family, encoding low-molecular-weight fatty acid binding proteins (FABP) are clustered. Hepatic lipase (HL), LPL, and pancreatic lipase also are members of one gene family but HL and LPL are unlinked. The genes for cholesterol ester transport protein (CETP) and LCAT reside in close proximity on chromosome 16 (q21 and q22, respectively) although they show no obvious homology. There is no indication that various

**Table 5-3. Chromosomes Location of
Genes of Plasma Apoproteins and Enzymes
Involved in Lipoprotein Metabolism**[a]

Gene	Chromosome location
Apo A-I	11q13-qter
Apo A-II	1p21-qter
Apo A-IV	11q13-qter
Apo B	2p23-p24
Apo C-I	19q
Apo C-II	19q
Apo C-III	11q-13qter
Apo D	3p14.2-qter
Apo E	19q
CETP	16q12-16q21
LDL receptor	19p
LPL	8p22
LCAT	16q22
HL	15q21
HMGR	5q13.1-q14
HMG-CoA synthase	5q14-p12

[a]From Lusis, 1988. Reproduced with permission.

cholesterol-regulated genes, including those for the LDL receptor, HMGR, or HMG-CoA synthase, are clustered.

5.2.1. ApoA-I

ApoA-I (mol. wts. 28,000–28,500) is the major protein of nascent chylomicrons and HDL and is an activator of LCAT. In humans, variants of apoA-I are poor activators of LCAT. ApoA-I also may play a role in reverse cholesterol transport (transport of cholesterol from peripheral tissues to liver). Plasma apoA-I levels in humans are increased by modest alcohol intake. The gene for this apoprotein occurs in a tight cluster (the apoA-I-C-III-A-IV gene cluster) spanning about 15 kilobases on the long arm of human chromosome 11.

5.2.2. ApoA-II

ApoA-II (mol. wt. 17,500) is the second most abundant protein of nascent chylomicrons and HDL; it is also an activator of LCAT and possibly HL. Like other apoproteins, it binds lipids. The protein has a high degree of ordered secondary structure, including amphipathic regions. The level of apoA-II in humans is increased by modest alcohol intake. In mice, the gene for apoA-II occurs on chromosome 1.

5.2.3. Apoa-IV

Apoa-IV (mol. wt. 46,000) occurs in nascent chylomicrons and HDL. In humans, it occurs in plasma to a large extent in the free form but can redistribute readily between the HDL and chylomicrons. Apoa-IV exhibits genetic polymorphism in humans. In humans the mRNA for apoa-IV is expressed only in intestines whereas in rats it is expressed in both liver and intestines. The gene for apoa-IV occurs on the apoa-I-C-III-A-IV gene cluster on chromosome 11.

5.2.4. Apob-100 and Apob-48

In humans, apob-100 occurs on VLDL and IDL, whereas apob-48 occurs in chylomicrons and chylomicron remnants. Apob-100 is highly insoluble and forms aggregates. It is synthesized in the liver and is essential for packaging the lipids into VLDL. Apob-100 is recognized by the LDL receptor and initiates the uptake of LDL into cells. The LDL receptor also recognizes apoE but not apob-48. When either lysine or arginine residues of apob-100 or apoE are chemically modified, the binding of LDL and IDL to the receptor is abolished. In humans, apob-48 is made in the intestines, whereas apob-100 is made in the liver. Apob-100 contains 4563 amino acids, whereas apob-48 contains only the amino-terminal 2152 amino acids of apob-100 (apob-48 is 48% as large as apob-100). Apob-48 lacks the carboxyl-terminal domain of apob-100 and thus does not bind to the LDL receptor. In chylomicron remnants, receptor binding is mediated by apoE. Schematic models of the apoproteins of chylomicrons, LDL, and Lp(a) are shown in Fig. 5-3.

Because of its large size and insolubility in water, the structure and amino acid sequence of apob-100 have been difficult to determine. The primary structure of the carboxyl-terminal end (1455 amino acids) has been deduced from the nucleotide sequence of the cDNA by Knott et al. (1985). The apob-100 mRNA is about 19 kilobases in length, and the gene is expressed primarily in the liver. The gene is located in the p24 region of chromosome 2. Most investigators agree that apob-100 is a single large polypeptide with a molecular weight about 550,000. Knott et al. (1986) and Yang et al. (1986) have determined the complete sequence of human apob-100 from its cDNA. Apob-100 has 4563 amino acids and a molecular weight of 514,000. Yang et al. (1986), by using synthetic peptides of a specific region of apob-100, have identified a potential LDL receptor binding domain at residues 3345–3381.

There is evidence for at least five allelic variations of apob-100 in humans. These were identified by use of alloantisera restriction endonucleases. The total number of common apob alleles is expected to be more than 14, since only a fraction of the genetic differences is likely to be revealed with alloantisera.

Apob-100 contains carbohydrate as 8–10% of its mass. Both mannose-rich chains and complex oligosaccharides that contain N-acetylglucosamine, galactose, mannose, and N-acetylneuraminic acid are found in the protein.

In the rare disease abetalipoproteinemia, apoB-48 and apoB-100 are absent from the blood, suggesting that the two proteins are made from one gene, possibly through alternative splicing of the mRNA. However, the cloned gene encoding apoB-100 shows no evidence for alternative splicing. Rather, the position at which apoB-48 terminates lies in the middle of a large exon of 7572 base pairs, the largest in any gene studied.

Powell et al. (1987) and Chen et al. (1987) have suggested a solution to this enigma. They used reverse transcriptase to clone 22 apoB-48 cDNAs from the intestines of three humans and one rabbit. In all of these clones, the mRNA was reverse transcribed as though it contained a uracil instead of the expected cytosine at position 6457 in the nucleotide sequence (relative to initiator codon AUG at +1). This substitution changes the codon 2153 from CAA (glutamine) to the termination codon UAA (Fig. 5-3). Oligonucleotide hybridization to genomic DNA confirmed a cytosine at nucleotide 6457 in the apoB gene of intestinal cells, just as in liver cells. Direct sequencing of the intestinal apoB mRNA demonstrated that this cytosine is replaced by a uracil during or after transcription. These data suggest that intestinal cells may possess a highly specific enzyme that modifies a single nucleotide position, 6457, in a single mRNA. The simplest mechanism would be deamination at the 6 position of cytosine, converting cytosine to uracil. This finding challenges the existing dogma that once an mRNA is formed, its coding region must remain intact or else chaos may result. The discovery of a sense–nonsense conversion in the mRNA for apoB raises the possibility of a new type of genetic editing.

5.2.5. ApoC-I

ApoC-I (mol. wt. 6600) is a single polypeptide consisting of 57 amino acids. The amino acid sequence was determined by protein sequencing and confirmed by nucleotide sequencing of cDNA. ApoC-I is an activator of LPL. The gene for apoC-I is located on the proximal region of the long arm of chromosome 19 and is part of the apoE-C-I-C-II gene complex.

5.2.6. ApoC-II

ApoC-II (mol. wt. 8824) is a single polypeptide consisting of 79 amino acids. Its amino acid sequence has been determined from its cDNA. It is an activator of LPL. The gene for apoC-II is located on chromosome 19 and is part of the apoE-C-I-C-II gene complex.

5.2.7. ApoC-III

ApoC-III (mol. wt. 8750) is a single polypeptide consisting of 79 amino acids. There are three forms of apoC-III (C-III0, C-III1, and C-III2), which differ in level of sialylation (number of sialic acid residues). ApoC-III may also be an activator of LPL. The gene for apoC-III is part of the apoA-I-C-III-A-IV gene cluster on chromosome 11 and occurs within three kilobases of the gene for apoA-I.

5.2.8. ApoE

ApoE (mol. wt. 34145) is a glycoprotein found in chylomicrons, chylomicron remnants, VLDL, IDL, and HDL. ApoE is a single polypeptide chain consisting of 299 amino acids. The human apoE gene is part of the apoE-C-I-C-II gene cluster located on chromosome 19. ApoE is synthesized as a preprotein having 317 amino acids. The preprotein undergoes intracellular proteolysis, glycosylation, and extracellular disialylation. ApoE binds to the LDL (or apoB/E) receptor. Whether there is a specific apoE receptor remains to be determined. ApoE appears to be synthesized in many organs since the mRNA for apoE has been detected in liver, brain, spleen, lung, adrenal gland, ovary, kidney, and muscle in several different species. The largest amount of apoE mRNA occurs in liver. ApoE produced in liver becomes a component of VLDL.

Like other apoproteins, ApoE has a high degree of ordered structure. particularly α-helical structure. The lipid binding domain on apoE is believed to occur in the carboxyl-terminal third of the polypeptide chain. The predicted secondary structure of apoE, showing α-helices, β-sheet, and β-turns is presented in Fig. 5-4.

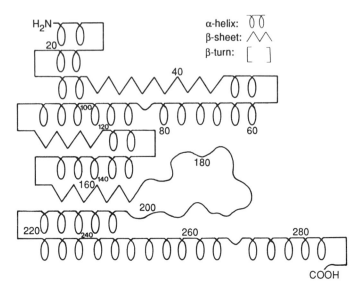

Figure 5-4. Schematic model of the structures of LDL (A) and the LDL receptor(B). (From Brown and Goldstein, 1986. Reproduced with permission.)

In plasma, three major isoforms, apoE4, apoE3, and apoE2, have been detected. These isoforms are determined by three alleles at a single gene locus. Other minor isoforms also exist and vary in degree of sialyation. The most common apoE allele is E3; the less common ones are E4 and E2. ApoE4 differs from apoE3 by an arginine-for-cysteine substitution at residue at 112. ApoE2 differs from apoE3 by a cysteine-for-arginine substitution at residue 158. As a result of these three alleles, six apoE phenotypes occur in the normal population. Three homozygous (E4/4, E3/3, and E2/2) and three heterozygous (E4/3, E4/2, and E3/2) phenotypes have been identified (Table 5-4). Isoelectric focusing gels of VLDL apoproteins with phenotypes apoE3/2, apoE3/3, apoE4/3, and apoE5/3 are shown in Fig. 5-5.

The most common phenotype is E3/3, which is present in about 60% of patients studied. The homozygous phenotype E2/2 has been shown to be predictive of occurrence of dysbetalipoproteinemia and is necessary for the expression of type III hyperlipoproteinemia (broad β disease). Type III disease may also be associated with other rare mutations of apoE and with apoE deficiency. ApoE4 now appears to be involved in type V hyperlipoproteinemia (Kuusi et al., 1988).

The levels of apoE lipoproteins are increased in the plasma of some animals after intake of diets high in fat and cholesterol; apoE becomes a major protein of two cholesterol-rich lipoproteins, β-VLDL and HDLc. β-VLDL may be atherogenic, since they enhance the massive accumulation of cholesterol esters in macrophages.

Recent studies by Bradley and Gianturco (1986) have shown that apoE is necessary and sufficient for the binding of large triglyceride-rich [hypertriglyceridemic (HTG)] lipoproteins (HTG-VLDL) from hypertriglyceridemic patients to the LDL receptor whereas apoB is not necessary. They found that the large VLDL from patients with hypertriglyceridemia, but not the VLDL from normal persons, bind to the LDL receptor because the large HTG-VLDL contain

Table 5-4. Prevalence of ApoE Phenotypes[a]

Phenotype	Prevalence (%)	
	Utermann et al.	Menzel et al.
E4/4	3	2
E3/3	60	63
E2/2	1	1
E4/2	23	20
E4/2	2	3
E3/2	12	11
Number of subjects	1031	1000

[a]From Mahley, 1988. Reproduced with permission.

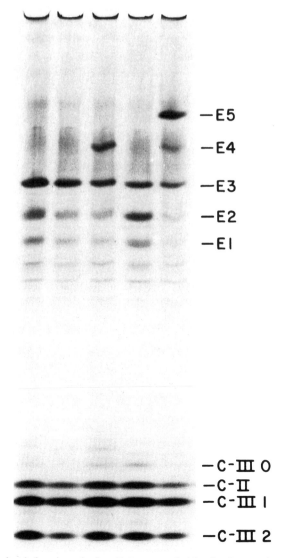

Figure 5-5. Isoelectric focusing gels of apoE phenotypes showing the pheenotypic patterns apoE3/2, apoE3/3, apoE4/3, apoE3/2, and apoE5/3. (From Ordovas, 1987. Reproduced with permission.)

apoE in a correct conformation. Thus, it appears that apoE is necessary for the binding of chylomicron remnants and large triglyceride-rich VLDL molecules to the LDL receptor.

The receptor-binding domain of apoE (Fig. 5-6) has been mapped in detail. Initially, it was established that a limited number of arginine and lysine residues

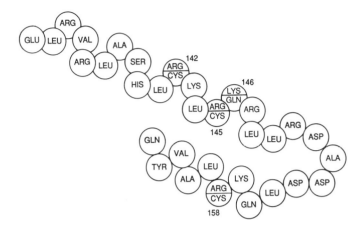

Figure 5-6. Receptor binding domain of apoE. (From Mahley, 1988. Reproduced with permission.)

within apoE (and also apoB-100) were essential for binding to the LDL receptor, since chemical modification of either one completely inhibited the binding of apoE and apoB-100 to the LDL receptor. The specific amino acid residues in apoE involved in mediating receptor binding have been identified by using four complementary experimental approaches: (1) identifying and sequencing natural apoE mutants defective in receptor binding, (2) generating apoE fragments and testing their binding activities, (3) mapping the epitope of an apoE monoclonal antibody that blocked binding of apoE-containing lipoproteins, and (4) producing site-directed mutant forms of apoE.

The most common variant of apoE associated with type III hyperlipoproteinemia is apoE2, in which cysteine replaces the normally occurring arginine at residue 158. However, several other rare apoE variants associated with this disorder also bind defectively. Sequencing studies showed that single amino acid substitutions in the defective mutants are clustered near residues 140–160. In all of these natural mutants, neutral amino acids substitute for the basic residue arginine or lysine. Fragments of apoE were produced by thrombin and cyanogen bromide treatment. Thrombin produced two major fragments, the amino-terminal residues 1–191 and the carboxyl-terminal residues 216–299. Only the amino-terminal fragment had binding activity. The only cyanogen bromide fragment with receptor activity contained residues 126–218. A third line of evidence also highlighted this same region of apoE. The epitope of a monoclonal antibody to apoE that blocked receptor binding was localized to residues 140–150.

The receptor-binding region of apoE is rich in arginine and lysine, which occur in doublets or triplets. The region from residue 131–150, predicted to be an α-helix, contains three sites at which substitutions disrupt receptor binding.

Residues 151–154 are predicted to form a β-turn. They are followed by a β-sheet containing residues 155–164. The variant at residue 158 appears to alter binding by altering the local molecular conformation.

The role of other specific amino acid residues in receptor binding has been elucidated by site-directed mutagenesis. Recombinant techniques can be used to produce an apoE in *E. coli* that displays normal binding and plasma clearance. These techniques were used to produce apoE mutants with alterations at specific sites in this region. Basic amino acids converted to neutral amino acids reduced binding to approximately 10–50% of normal, about the same range as is obtained with naturally occurring variants. The substitution of proline for leucine-144 or for alanine-152 decreased binding significantly. However, a double mutation at both residues 139 and 149 not only failed to decrease receptor binding but apparently increased binding.

These studies show that the basic amino acids in the vicinity of residues 140–160 are important in mediating the binding of apoE to the LDL receptor. The molecular conformation in this region is important for normal receptor binding, but substitutions outside this immediate region also could have an effect on the binding domain.

ApoE also appears to be involved in the repair response to tissue injury since markedly increased amounts of apoE are found at sites of peripheral nerve injury and regeneration. Other functions of apoE unrelated to lipid transport appear to include immunoregulation and modulation of cell growth and differentiation.

This information on apoE was taken from a review article by Mahley (1988).

5.3. METABOLISM OF LIPOPROTEINS

5.3.1. Lipoprotein Lipase

LPL has a major role in the metabolism of lipoproteins. Its primary function is to catalyze the hydrolysis of the core triacylglycerols of circulating chylomicrons, VLDL, and IDL. During this hydrolysis, the chylomicrons, VLDL, and IDL are converted to smaller particles that enable the transfer of cholesterol esters on these particles to various tissues of the body. LPL is synthesized in parenchymal cells of several tissues but functions on the luminal surface of the vascular endothelium, where it is anchored by a membrane-bound glycosylaminoglycan-phosphatidylinositol chain. Recent work (Chan et al., 1988) has indicated that insulin stimulates the release of LPL into the blood by enhancing a phospholipase C that hydrolyzes the phosphatidylinositol anchor. In the presence of the cofactor apoC-II, LPL hydrolyzes the triacylglycerols mainly to free fatty acids and glycerol. Some monoacylglycerols and diacylglycerols may also occur if the hydrolysis is not complete. The fatty acids are taken up by cells for oxidation or for storage. Oxidation of the fatty acids occurs in most cells, especially those of muscle, heart, kidney, and liver. The fatty acids are stored as triacylglycerols in adipose tissue.

A cDNA for human LPL has been cloned and sequenced. This cDNA codes for a mature protein of 448 amino acids. Analysis of the sequence indicates that human LPL, HL, and pancreatic lipase are members of one gene family. The gene for LPL occurs on chromosome 8 in humans.

5.3.2. Hepatic Lipase

HL appears to function primarily in the metabolism of cholesterol ester-rich lipoproteins. The enzyme is localized on the sinusoidal surfaces of the liver and, like LPL, can be released by the injection of heparin. The cDNA for HL has been isolated and sequenced and found to be a member of a dispersed gene family of lipases that include LPL and pancreatic lipase. Rare deficiencies of HL are characterized by abnormal triacylglycerol-rich LDL and HDL and the presence of β-VLDL.

5.3.3. Metabolism of Chylomicrons

The exogenous triacylglycerol-rich lipoproteins, called chylomicrons, are the largest lipoproteins (100 nm in diameter). They are synthesized in the small intestine and carry primarily triacylglycerols and some cholesterol via the lymphatics to the blood. A schematic representation of this process is shown in Fig. 5-7.

Circulating chylomicrons contain apoproteins A-I, A-IV, B-48, C, and E. The newly synthesized (nascent) chylomicrons probably contain only apoA and apoB-48 and acquire the other apoproteins in the plasma either directly or by exchange with HDL. Because of these protein exchange reactions, it is difficult to determine the absolute protein composition of the lipoproteins. Chylomicrons are degraded in the blood by the enzyme LPL, which hydrolyzes the tri-acylglycerols to free fatty acids and glycerol (and some monoacylglycerols) and converts the chylomicrons to smaller particles called chylomicron remnants, which contain apoB-48 and apoE and have lost apoC and possibly apoA. The fatty acids that are released can be taken up by most cells, but adipocytes take up the bulk and store them as triacylglycerols. The storage of fat occurs after food is consumed and is stimulated by insulin. Liver is the main organ that utilizes the glycerol since it contains glycerol kinase which converts glycerol to glycerolphosphate, which is then used for the synthesis of phospholipids and triacylglycerols. Adipocytes lack glycerol kinase and must derive glycerolphosp-

Intestinal lumen

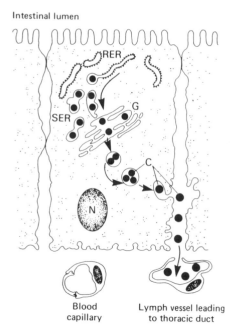

Figure 5-7. Formation and secretion of chylomicrons by intestinal mucosal cells. SER, Smooth endoplasmic reticulum; RER, rough endoplasmic reticulum; G, Golgi; C, chylomicrons; N, nucleus. (From Murray, et al., 1988. Reproduced with permission.)

Blood capillary

Lymph vessel leading to thoracic duct

hate from glycolysis. Insulin is important for glycolysis, since it enhances glucose uptake by fat cells.

The metabolism of chylomicrons in blood plasma is shown in Fig. 5-8. The chylomicron remnants are taken up by liver cells which have specific membrane receptors (the LDL or the apoB/E receptor), which bind apoB-100 and apoE. Some investigators believe that there is another specific apoE receptor. Apparently the function of apoB-48 is to allow for the effective binding of apoE to the apoB/E receptor. Apob-48 lacks the receptor-binding domain that is present in apoB-100. The chylomicron remnants are taken up by a process called receptor-mediated endocytosis.

Chylomicrons have a half-life in the blood about 15 min. For lipoprotein lipase to degrade chylomicrons, the chylomicrons must pick up apoC-II from the plasma or from HDL, since apoC-II is an important activator of this enzyme. HL is believed to degrade chylomicron remnants to smaller particles that are enriched in cholesterol, and some behave like HDL particles. Lysosomal proteases, lipases, and phosphodiesterases in liver cells degrade the intracellular chylomicron remnants to free amino acids, fatty acids, glycerol, N bases (choline, ethanolamine, and serine), and free cholesterol. The cholesterol is converted in large part to bile acids. The degradation products of chylomicron remnants also are used for the synthesis of VLDL and HDL.

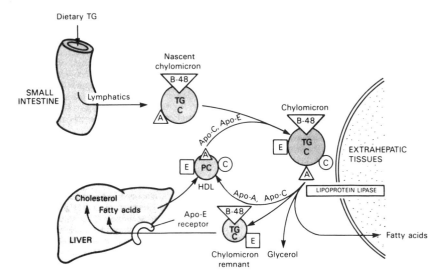

Figure 5-8. Metabolism of chylomicrons. TG, Triacylglycerols; C, cholesterol plus cholesterol esters; P, phospholipid. The triangles, small circle, and small square represent the apoprotins apoA, apoB-48, apoC, and apoE. (From Murray, et al., 1988. Reproduced with permission.)

Elevated levels of chylomicrons give the plasma a milky appearance and cause hypertriglyceridemia, since these lipoproteins are very high in triacylglycerols. Genetic defects in the synthesis of LPL give rise to a severe hypertriglyceridemia called hyperlipoproteinemia type Ia. A deficiency of apoC-II gives rise to hyperlipoproteinemia type Ib, in which plasma levels of triacylglycerols (as chylomicrons) are greatly elevated.

5.3.4. Metabolism of Very-Low-Density Lipoproteins

The endogenous lipoproteins named VLDL are produced in the liver. They transport triacylglycerols, phospholipids, and cholesterol from the liver to other tissues of the body. VLDL have a particle size of 30–90 nm and a density of $<1.006 \text{ g/ml}$. VLDL are degraded by LPL stepwise, first into IDL and then to LDL. IDL have a particle size about 20–30 nm and a density of 1.006–1.019 g/ml. LDL have a particle size <20 nm and a density of 1.019–1.063 g/ml. The synthesis of VLDL particles in a liver cell is depicted in Fig. 5-9.

The apoproteins of VLDL are synthesized in the rough endoplasmic reticulum (RER) in close proximity to the smooth endoplasmic reticulum (SER), where cholesterol, phospholipids, and triacylglycerols are produced. The lipids and apoproteins are assembled, packaged, and transported via secretory vesicles to the plasma membrane. The secretory vesicles fuse with the plasma membrane

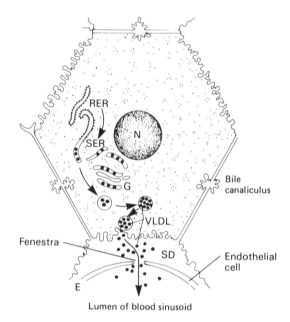

Figure 5-9. Synthesis and secretion of VLDL in liver. SER, Smooth endoplasmic reticulum; RER, rough endoplasmic reticulum; G, Golgi; N, nucleus; SD, space of Disse.

and release the VLDL into the space of Disse. The VLDL then migrate to the sinusoids and enter the blood. There may be some modification of the VLDL in the Golgi. This modification usually involves the addition or removal of carbohydrate residues on the apoproteins.

The nascent VLDL secreted in the blood contain apoB-100 and apoE but probably pick up apoC and additional apoE from the plasma or from HDL. Apoc-II activates LPL which degrades the triacylglycerols of VLDL, yielding IDL and then LDL (Fig. 5-10). During this stepwise degradation, apoC and apoE are released from the VLDL so that the IDL contain only apoB-100 and apoE and the LDL contain only apoB-100. The released apoC and apoE undergo recycling that is mediated by HDL and transfer proteins.

The liver and extrahepatic tissues contain the apoB/E receptors that recognize and bind apoE and apoB-100, permitting lipoprotein particles containing these apoproteins to undergo receptor-mediated endocytosis. Since IDL contain both apoB-100 and apoE, these particles are more rapidly taken up by liver and other cells than are LDL, which contain only apoB-100. ApoE has a higher affinity for the receptor than does apoB-100. Thus, IDL do not normally accumulate in the plasma, and consequently LDL represent the major degradation end product of VLDL metabolism. In type III hyperlipoproteinemia, however, there

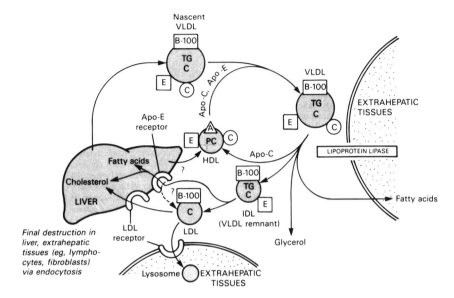

Figure 5-10. Metabolism of VLDL. TG, Triacylclycerols; C, cholesterol plus cholesterol esters; P, phospholipid. The small circle, triangle, square, and rectangle represent the apoproteins apoA, apoB-100, and apoE.

are genetic defects in apoE that do not permit the rapid uptake of these particles; as a result they accumulate in plasma and give the plasma a turbid appearance. The VLDL and IDL particles in type III persons are abnormal and have been called β-VLDL. Apparently, these particles are avidly taken up by macrophages and produce foam cells that are laden with cholesterol and cholesterol esters and stimulate the process of atherogenesis.

5.3.5. Metabolism of High-Density Lipoproteins

HDL represent several lipoprotein species (designated HDL-1, HDL-2, HDL-3, and HDL-4) that differ in their protein and lipid content, shape, structure, and density. There is progressive loss of phospholipids, cholesterol, and cholesterol esters when HDL are transformed from HDL-1 to HDL-4. The HDLs have a particle size of 8–12 nm and a density of 1.063–1.21 g/ml. They are made in the liver and, to some extent, in the small intestine. HDLs contain apoA, apoC, and apoE. ApoC and apoE are transferred between HDLs and VLDL and chylomicrons. Nascent HDL made in the liver (or HDL-like precursors, which are membrane fragments made in the plasma by the action of LPL and HL on VLDL and chylomicrons) are disklike particles containing mainly phospholipids,

free cholesterol, and proteins but very little cholesterol esters and triacylglycerols. The disks are transformed to spherical HDL particles by the addition of cholesterol esters via the action of the enzyme LCAT. Some cholesterol
esters are then transferred to other lipoproteins such as VLDL, IDL, and LDL.
LCAT is activated by apoA-I, apoA-II, and apoA-IV and possibly by apoC-I.
LCAT transfers an unsaturated fatty acid from the 2 position of lecithin (phosphatidylcholine) to the hydroxyl group of cholesterol, producing cholesterol ester
and lysolecithin (lysophosphatidylcholine). During this process, the HDL particle undergoes a change in shape from disk to sphere. The core of the HDL sphere
contains cholesterol esters.

The transfer of cholesterol esters is believed to require a specific CETP. The
metabolism of HDL is shown in Fig. 5-11.

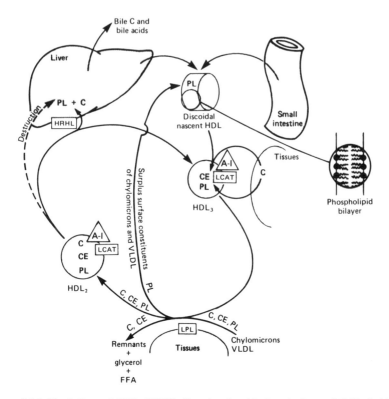

Figure 5-11. Metabolism of HDL. HRLH, Heparin-releasable hepatic lipase; LCAT, lecithin-
cholesterol acyltransferase; LPL, lipoprotein lipase; C, cholesterol; CE, cholesterol esters; PL,
phospholipids, FFA, free fatty acids; A-I, apoprotein A-I. The triangle represents apoA-I but may
also contain apoA-II, apoA-III, and apoA-IV. (From Murray, et al., 1988. Reproduced with permission.)

When HDL transfer cholesterol esters to VLDL and LDL, the HDL lose cholesterol and thus are able to pick up additional cholesterol from peripheral tissues and transport them to liver, where the cholesterol can be converted to bile acids. This process, called reverse cholesterol transport, is believed to explain why high levels of HDL offer protection from coronary heart disease. Studies on macrophages (Aviram et al., 1989) indicate that binding of HDL to its receptor on cholesterol-loaded macrophages appears to promote translocation of intracellular cholesterol to the plasma membrane, followed by cholesterol efflux into the medium. However, in nonloaded macrophages, HDL stimulates cholesterol movement from the plasma membrane into intracellular pools by a receptor-independent pathway.

HL, whose release into the blood is stimulated by heparin, is believed to degrade phospholipids on the surface of HDL-2. The presence of apoE on HDL targets this lipoprotein to cells containing the apoB/E receptor. The transport of cholesterol from peripheral tissues to the liver is the major way for the body to eliminate excess cholesterol, since only the liver can convert cholesterol to bile acids. It is believed that as HDL pick up cholesterol from peripheral tissues and become enriched in cholesterol, they acquire apoE, which is available in interstitial fluid as a product secreted by many cells, including macrophages and smooth muscle cells. ApoE has an avidity for HDL and facilitates the acquisition of cholesterol by HDL. As the HDL become enriched in cholesterol they change shape and increase in diameter. The enzyme LCAT then converts cholesterol to cholesterol esters which are arranged in concentric layers in the core of the particle. These particles bind avidly to the LDL (apoB/E) receptor by virtue of the presence of apoE. HDL lacking apoE do not bind to the LDL receptor. The cholesterol esters of HDL also are transferred in part to VLDL, IDL, and LDL, which are taken up by the liver via receptor-mediated endocytosis.

It is likely that HDL enriched in apoE can function in the delivery of cholesterol to cells in the environment in which they are formed. Thus, apoE could participate in the redistribution of cholesterol from cells with excess cholesterol to other cells requiring cholesterol.

A possible role of apoE in injured and regenerating peripheral nerves has been proposed. After a crush or cut injury to the rat sciatic nerve, apoE is produced and accumulates to levels 100–200 fold greater than in uninjured nerve. The extracellular apoE can attain levels representing 5% of the total soluble protein in the regenerating nerve segment. The cell responsible for apoE synthesis in this situation is the macrophage. Monocytes rapidly enter the site of injury, become macrophages, and secrete apoE. This nerve model may describe a more general process that occurs to a greater or lesser extent in various tissues in response to injury and repair. For a review of this topic, see Mahley (1988).

To indicate how the various lipoproteins are interrelated, an overview of the metabolism of lipoproteins is given in Fig. 5-12. The figure portrays that

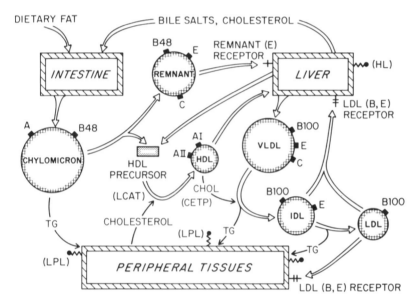

Figure 5-12. Overview of the metabolism of plasma lipoproteins. AI, AII, B48, B100, C, and E represent the apoproteins that occur on the surface of the indicated lipoproteins. (From Lusis, 1988. Reproduced with permission.)

plasma lipoproteins are dynamic macromolecules in a constant state of synthesis and degradation. They act to transport lipids from various tissues to the liver and from the liver to peripheral tissues of the body. Lipids that are digested and absorbed in the gut lead to the formation of chylomicrons rich in triacylglycerols and containing apoA and apoB-48. ApoC and apoE are picked up by chylomicrons in the plasma. This lipoprotein particle is altered by LPL, which is attached to the endothelium of capillaries. LPL hydrolyzes triacylglycerols of chylomicrons to glycerol and free fatty acids, which are taken up by many cells, particularly adipose tissue cells. The resulting smaller chylomicron remnants are then taken up rapidly by the liver through receptor-mediated endocytosis.

The liver secretes some cholesterol and bile acids via the biliary system to the gut. This creates an enterohepatic circulation of sterol and bile acids and a pathway for elimination of sterol metabolities via the feces. The liver also secretes VLDL, whose components may be derived either from the diet or by endogenous biosynthesis. The rate of secretion of VLDL is variable, depending on genetic and environmental conditions. VLDL contain apoB-100, apoC, and apoE and also serve as a substrate for LPL which converts VLDL to IDL. IDL have two fates, uptake by the liver via LDL receptors and conversion to LDL. LDL, the principal form of lipoprotein in human plasma, are taken up by LDL receptors in both liver and peripheral tissues. A small amount of LDL is also taken up by a non-receptor-mediated pathway in reticuloendothelial and arterial endothelial cells. HDL are made in the liver from lipids and apoA, secreted into plasma as a discoidal form, and modified in plasma by binding apoE and by LCAT. HDL playss an important role in reverse cholesterol transport bringing excess cholesterol from peripheral tissues to the liver. LCAT, HL, and CETP are involved in this process.

5.4. LDL, LDL RECEPTOR, AND HMGR

5.4.1. Metabolism of LDL

Animal cells have specific mechanisms for obtaining the cholesterol required for synthesis of cell membranes, steroid hormones, and bile acids. Cells can synthesis cholesterol de novo, or they can take up dietary cholesterol from the blood by receptor-mediated endocytosis utilizing the cholesterol-rich lipoproteins LDL, IDL, and chylomicron remnants. The metabolism of LDL via receptor-mediated endocytosis is shown in Fig. 5-13. This schematic diagram is called the Brown and Goldstein model. The LDL particle must first bind to a specific receptor, the LDL (apoB/E) receptor on the cell plasma membrane. There are approximately 40,000 receptors per cell. Adrenal gland and liver have more receptors per cell than do other cells. The receptors are localized in discrete regions of the membrane called coated pits. The coated pits make up about 2% of the cell surface and contain the protein clathrin, which consists of heavy chains (mol. wt. 192,000) and light chains (mol. wt. 32,000–38,000). These proteins form a fibrous network of 12 pentagons and 8 hexagons in the coated pits. The fibrous network is constructed of 36 clathrin triskelions. Polymerization of clathrin into a lattice along the cytoplasmic (inner) surface of the plasma membrane is believed to cause the pit to expand and pinch off from the membrane after the LDL has bound to the receptor.

The internalization of the LDL occurs in several steps and begins within 2–5 min after binding. During uptake, the coated pit pinches off and becomes a coated vesicle. The clathrin lattice depolymerizes and forms an uncoated vesicle called an endosome. The endosome is transformed into an uncoupling vesicle called a curl (compartment of uncoupling of the receptor and ligand), which has a low internal pH of about 5.0. The low pH, which is maintained by an ATP-driven pump, causes the LDL to dissociate from the receptors. A receptor-rich region buds off to form a recycling vesicle that recycles a major part of the receptors back to the plasma membrane. The endosome containing the LDL fuses with a lysosome forming a larger secondary lysosome, where the apoB and apoE proteins are degraded to amino acids by proteases and the cholesterol esters

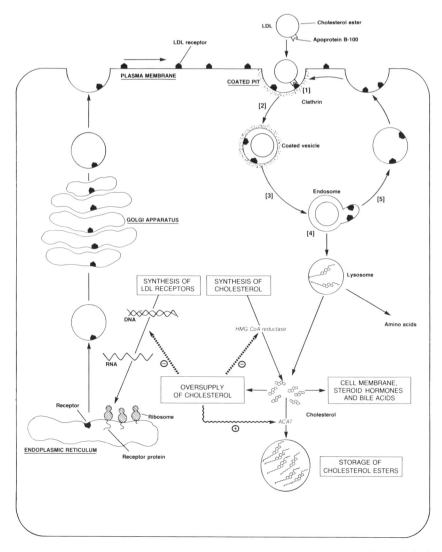

Figure 5-13. Metabolism of LDL. (From Champe and Harvey, 1987. Reproduced with permission.)

are hydrolyzed to free cholesterol and fatty acids by a cholesterol esterase. The triacylglycerols are hydrolyzed to free fatty acids and glycerol by lipases. Either the free cholesterol is incorporated into the various cell membranes or the excess is transported out of the cell to the liver by HDL-mediated reverse cholesterol transport. The free fatty acids can be oxidized to furnish energy to the cell. In rapidly dividing cells, the cholesterol and phospholipids are required to make new membranes. The synthesis of new membranes also requires the synthesis of specific proteins that reutilize the degradation products of LDL. Although most cells can make cholesterol from acetyl-CoA, the liver is the major organ that performs this function. Nonhepatic cells derive most of their cholesterol from serum lipoproteins made in the liver (i.e., VLDL, which are then converted to LDL). As the LDL are taken up by these cells and cholesterol accumulates in the cell, the de novo synthesis of cholesterol decreases or stops. The uptake of LDL by cells is highly regulated. This regulation occurs by four mechanisms. As the cholesterol level in the cell increases, synthesis of the key regulatory enzyme HMGR is inhibited. Second, there is inhibition of the synthesis of the enzyme cholesterol esterase, which hydrolyzes cholesterol esters to free cholesterol and fatty acids. Third, synthesis of the enzyme acyl-CoA cholesterol acyl trans-ferase, which converts free cholesterol to cholesterol esters, is stimulated. Fourth, inhibition of the synthesis of LDL receptors develops. This inhibition is believed to result from the formation of a protein that inhibits the synthesis of mRNA coding for the LDL receptor. The regulation of receptor synthesis allows the cells to take up only the amount of cholesterol needed.

Cells from individuals with genetic defects in the LDL receptor cannot obtain sufficient cholesterol from blood lipoproteins (mainly LDL) and therefore produce more cholesterol de novo via an increased synthesis of HMGR. In normal cells, the LDL receptor and HMGR are under feedback regulation so that when cholesterol levels in the cell rise too high, syntheses of the LDL receptor and HMGR are repressed. If this balance is not maintained, hyper-cholesterolemia will occur. When blood levels of LDL fall too low, activities of the LDL receptor and HMGR are increased via an increased synthesis of their mRNAs.

5.4.2. Structure of LDL and the LDL Receptor

The LDL receptor was purified from bovine adrenal gland by Schneider et al. (1982). A partial amino acid sequence was later used by other researchers to isolate a full-length cDNA for the human LDL receptor. Schematic models of the LDL receptor and LDL are shown in Fig. 5-14.

The LDL receptor consists of 839 amino acids. The protein can be divided into five domains. The first domain consists of 292–322 amino acids at the N-terminal end and is rich in disulfide bonds. It is very rigid, contains multiple loops, and confers stability to the protein. This domain contains clusters of negatively charged amino acids that are believed to bind to the positively charged lysine and arginine residues of apoB and apoE. It therefore is the binding domain for LDL, IDL, and chylomicron remnants. The second domain contains 350–400 amino acids that have a 35% homology with a portion of the extracellular

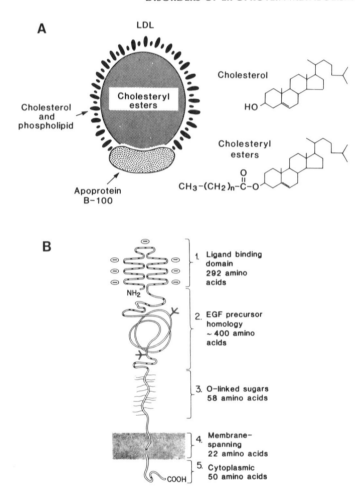

Figure 5-14. Schematic model of the structures of LDL (A) and the LDL receptor (B). From Brown and Goldstein, 1986. Reproduced with permission.)

domain of the protein precursor of the epidermal growth factor (EGF). This finding has led to speculation that the LDL receptor evolved from the same gene that gave rise to EGF, a protein that stimulates growth.

The third domain consists of 48–58 amino acids of which 18 are serine and threonine residues containing carbohydrate molecules linked by O-glycosidic bonds. This domain is located just external to the plasma membrane and may serve to extend the LDL receptor so that the binding sites for apoB and apoE are available to the lipoproteins containing these apoproteins. The fourth domain has a membrane-spanning region of 22 amino acids and acts to anchor the receptor in the membrane. The fifth domain, at the carboxyl end, contains 50 amino acids and projects into the cytoplasm. These sequences are conserved in the human and bovine receptors and are believed to bind to clathrin. Binding to clathrin initiates the process of endocytosis, which allows the cell to take up lipoproteins from the blood.

LDL is a spherical particle with a mass of 3×10^6 daltons and a diameter of 22 nm (Fig. 5-14B). Each LDL particle contains about 1500 molecules of cholesterol ester in an oily core that is shielded from the aqueous plasma by a hydrophilic, polar membrane composed of 800 molecules of phospholipid, 500 molecules of unesterified cholesterol, and 1 molecule of apoB-100. Elevations in blood cholesterol are usually attributable to an increase in the number of LDL particles.

5.4.3. Variant of the LDL

An interesting article by Brown and Goldstein (1987) discusses a variant of LDL called Lp(a) which was discovered by Berg in 1963 as an antigen in the blood of certain individuals. Lp(a) is an LDL that carries one or two copies of a protein called apo(a) joined to apoB-100 by a disulfide bond (Fig. 5-3).

Intestinal and hepatic apoB differ in size and function. ApoB-100, made in the liver and found in LDL and VLDL, contains 4563 amino acids whereas apoB-48, made in the intestine and found in chylomicrons, contains only the amino-terminal 2152 amino acids of apoB-100 (apoB-48 is 48% as large as apoB-100). ApoB-48, which lacks the carboxyl-terminal domain of apoB-100, does not bind to the LDL receptor. In chylomicrons remnants, receptor binding appears to be mediated mainly by apoE.

The amount of Lp(a) in plasma varies from undetectable to 100 mg/dl. High levels of Lp(a) are strongly associated with atherosclerosis. When the level is above 30 mg/dl, as prevails in about 20% of people, the relative risk of coronary atherosclerosis rises twofold. When LDL and Lp(a) are both elevated, the risk is about fivefold higher.

The mechanism by which Lp(a) accelerates atherosclerosis is not understood, but an interesting clue has emerged from recent work. Through protein sequencing and cDNA cloning, it has been found that apo(a) is a deformed relative of plasminogen, the precursor of the proteolytic enzyme plasmin, which dissolves fibrin clots. Plasminogen, a protein of 791 amino acids, contains five cysteine-rich sequences of 80–114 amino acids. These sequences are called kringles. Each kringle contains three internal disulfide bridges producing a pretzel-like structure that resembles a Danish cake called kringle. These kringles also are found in other proteases of the coagulation system, including tissue plasminogen activator (TPA) and prothrombin. In plasminogen, the kringles promote the binding to the substrate fibrin. Plasminogen is inactive until it has been cleaved by TPA, which breaks the chain at a single arginine residue to form active plasmin. Apo(a) contains a hydrophobic signal sequence for secretion followed by 37 copies of kringle 4 of plasminogen and then by kringle 5 and the protease domain. The 36th copy of a kringle 4 is altered to contain an extra unpaired cysteine, the likely site of disulfide linkage with apoB-100. Apo(a) is not a protease because the arginine at the cleavage site for TPA is changed to serine.

These findings may provide new insight into the interaction between plasma lipoproteins and the clotting system. There is evidence that microthrombi on the blood vessel wall can become incorporated into the atherosclerotic plaque. Fibrinogen and fibrin are found in plaques in amounts roughly equal to that of cholesterol. Lp(a) is also present in plaques. Kringle 4 binds to fibrinogen, although weakly. A protein with 37 copies of this kringle might bind to fibrinogen or fibrin under certain conditions in vivo, although no such binding has yet been detected in vitro. When Lp(a) insinuates itself into the arterial wall following endothelial damage, it may adhere to fibrin forming a complex that becomes stuck to the wall. Lp(a) may also block proteolysis of fibrin by inhibiting the activation of plasminogen by TPA and may also compete with plasminogen for access to fibrin, thereby inhibiting the dissolution of fibrin clots.

The frequency of high apo(a) levels is about the same among Japanese people as it is in Caucasians, yet atherosclerosis in Japan was much less frequent than in the United States until the Japanese began to consume a high-cholesterol diet and developed elevated plasma LDL levels. Armstrong et al. (1986) have demonstrated an additive effect between LDL and Lp(a) in producing angiographically detectable coronary artery disease.

Drugs that lower LDL levels may alleviate the toxic effects of Lp(a) without affecting the level of the lipoprotein. Bile acid-binding resins such as colestid, which lower LDL levels by increasing the number of LDL receptors, do not affect the levels of Lp(a). Other drugs, such as nicotinic acid, that lower LDL by other mechanisms do lower the level of Lp(a).

These studies demonstrate the impact of modern methods of molecular and cell biology in understanding the genetic and biochemical basis of protein structure and function and in elucidating the molecular basis of diseases.

5.4.4. Synthesis of the LDL Receptor

The LDL receptor is synthesized in the RER as a precursor (mol. wt. 120,000) that contains mannose-rich N-linked carbohydrate chains and the core sugar N-acetylgalactosamine of the O-linked chains. The O-linked core sugars are added before the mannose residues of the N-linked chains are trimmed. Within 30 min after synthesis, modification of carbohydrate residues occurs. The precursor contains at least two mannose-rich N-linked oligosaccharides that are converted in the Golgi to more complex N-linked oligosaccharides by addition of galactose and neuraminic acid (sialic acid) residues. About 45 min after synthesis, the LDL receptors appear on the cell surface plasma membrane, where they gather in coated pits. It is estimated that each LDL receptor makes one round trip every 10 min. The signals that regulate receptor cycling remain to be elucidated.

5.4.5. The LDL Receptor mRNA and Gene

The sequence of the cDNA for the human LDL receptor reveals that the protein contains a leader (signal) sequence of 21 amino acids that is cleaved before migration of the protein to the cell plasma membrane. Studies on cultured cells using hybridization techniques have shown that the mRNA is reduced in amount when cholesterol and other sterols such as hydroxycholesterol are added to the culture medium.

The structural gene for the human LDL receptor occurs on chromosome 19, a finding that agrees with family linkage data that place the locus responsible for familial hypercholesterolemia (FH) on chromosome 19. The gene cluster apoE-C-I-C-II is on the same chromosome. However, the gene for apoB-100 is located on chromosome 2 (Knott et al., 1985).

5.4.6. Mutations of the LDL Receptor

Many mutations in the LDL receptor have been found in patients with FH. These allelic mutations can be divided into four classes. In class 1, the most frequent form, the mutant gene either fails to synthesize mRNA or synthesizes such a faulty mRNA that the protein product cannot be identified by antibodies to the normal protein. Class 2 mutations allow the synthesis of receptors in the RER but the receptors are not transported to the Golgi and do not undergo carbohydrate processing. Therefore, these receptors do not get transported to the plasma membrane. Class 3 mutations enable the synthesis of receptors and their transport to the plasma membrane, but the receptors fail to bind LDL normally. Class 4 mutations are those in which the receptors are synthesized, processed, and transported to the cell membrane, where they bind LDL in a normal manner, but fail to cluster in the coated pits and are not internalized. This indicates that the defect is in the cytoplasmic carboxyl region of the receptor, which is responsible for binding to clathrin.

Recently, Yamamoto et al. (1986) cloned and sequenced cDNAs for the LDL receptor from normal and Watanabe heritable hyperlipidemic (WHHL) rabbits and found that the defect in the WHHL rabbits arose from an in-frame deletion of 12 nucleotides resulting in the elimination of four amino acids from the cytosine-rich binding domain. A similar mutation was found in a human with FH whose receptor failed to be transported to the cell plasma membrane. To date, nine mutations have been identified by molecular cloning and DNA sequence analysis or by restriction endonuclease analysis of genomic DNA. Very recent work (Soria et al., 1989) has revealed a mutation in the codon for amino acid 3500 resulting in a substitution of glutamine for arginine in the human apoB. This mutation CGG→CAG) occurs in the region of apoB that is associated with binding to the LDL receptor. Patients with this mutation have hypercholesterolemia.

Figure 5-15 depicts some mutations affecting the cytoplasmic domain of the LDL receptor in three patients who are homozygous for FH and have the internalization-defective form of this disease. The mutations can lead to a missense of the code (tyr→cys), a frameshift insertion of four bases, or a nonsense-type mutation (trp→stop codon).

5.4.7. Structure of HMGR

HMGR catalyzes the rate-limiting step in the synthesis of cholesterol. This enzyme occurs in the endoplasmic reticulum (ER). Cloning of its cDNA was made possible through the development of a line of Chinese hamster ovary cells (UT-1) that was adapted to growth in the presence of compactin. These cells had a 15-fold amplification of the gene for HMGR and transcribed the mRNA for HMGR at a 20-fold higher rate than did normal cells. HMGR from hamster has a molcular weight of 97,902 and consists of 887 amino acids. The protein has an N-terminal domain of 35,000 daltons that is very hydrophobic and is believed to crisscross the ER membrane seven times. The N-terminal end of this segment protrudes into the lumen of the ER. The carboxyl-terminal end is contiguous with a 62,000 dalton water-soluble stretch of amino acids that protrudes into the cytoplasm. This cytoplasmic domain contains the catalytic site of the enzyme.

5.4.8. Synthesis of HMGR

Synthesis of HMGR occurs on the RER. cDNA sequencing and cell-free translation studies show that HMGR does not contain a hydrophobic leader sequence at its N terminus. The enzyme contains at least one N-linked carbohydrate chain that does not undergo modification suggesting that HMGR does not enter the Golgi.

5.4.9. HMGR mRNA and Gene

In humans, the gene for HMGR is located on the long arm of chromosome 5; in mice, it is located on the distal end of chromosome 13. Many DNA

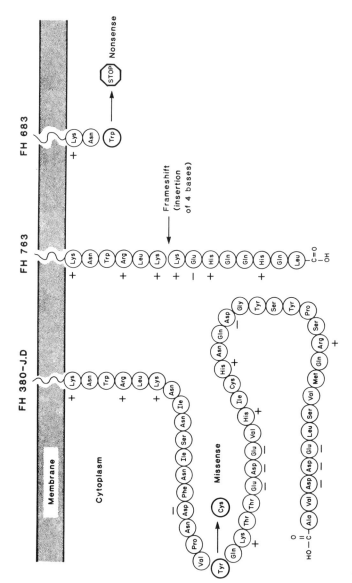

Figure 5-15. Mutations affecting the cytoplasmic domain of the LDL receptor in three FH homozygotes with internalization-defective form of FH. (From Brown and Goldstein, 1986. Reproduced with permission.)

polymorphisms of HMGR occur among inbred strains of mice. No DNA polymorphisms of the human HMGR gene have been reported. The gene for the hamster HMGR spans 25,000 nucleotides and contains 20 exons that are separated by 19 introns. Each of the seven hydrophobic segments of HMGR is specified by a separate exon. When added to culture media, cholesterol and other sterols such as 25-hydroxycholesterol suppress the activity of HMGR by suppressing the synthesis of the mRNA that codes for HMGR.

5.4.10. Coordinate Regulation of HMGR and the LDL Receptor

Cells are able to regulate their need for cholesterol. Two tissues, liver and adrenal gland, have a large requirement for cholesterol and express a large number of LDL receptors. The adrenal gland uses the abundant supply of LDL receptors to provide most of the cholesterol for synthesis of steroid hormones. On the other hand, the liver uses its LDL receptors to supply the cholesterol for synthesis of lipoproteins and bile acids. The levels of LDL receptors in livers of animals fed diets high in cholesterol are reduced, and cholesterol levels in plasma rises. Conversely, the number of LDL receptors is increased when the liver's demand for cholesterol is increased. This occurs when bile acids are shunted out of the body either by the use of bile acid-binding resins such as cholestryamine (colestid) or by draining of bile out of the small intestine.

5.5. PATHOGENESIS OF HYPERLIPOPROTEINEMIAS

Hyperlipoproteinemia represents the abnormal elevation of one or more of the plasma lipoproteins. Depending on the lipoprotein involved, this is reflected in elevated levels of plasma cholesterol, triacylglycerol, or both cholesterol and triacylglycerol. Beaumont and co-workers (1970) have arbitrarily classified the hyperlipoproteinemias into five types (indicated as I–V or 1–5). Some of these types are divided into subtypes (e.g., Ia, Ib, IIa, and IIb). The five major types are classified on the basis of electrophoretic patterns (Table 5-5).

Although this classification has been helpful in guiding physicians in the treatment of a particular disease, current research on the apoproteins of the various lipoproteins has led to new thinking on the pathogenesis of the hyperlipoproteinemias. It is likely that the protein abnormalities involved in these lipid disorders will lead to a new classification. The field of lipoprotein metabolism is moving very fast, and there is still much to learn about how the lipoproteins are synthesized and metabolized.

The term hyperlipoproteinemia is widely used. Other commonly used terms, such as hyperlipidemia, hypertriglyceridemia, and hypercholesterolemia, attempt to focus either on the particular lipoprotein or the particular lipid that is

Table 5-5. Types of Hyperlipoproteinemias

Type	Increased lipoprotein	Increased lipid[a]	Defect or deficiency	Fasting plasma appearance
Ia	Chylomicrons	TG	LPL	Milky
Ib	Chylomicrons	TG	Apoc-II	Milky
IIa	LDL	C	LDL receptor	Clear
IIb	LDL + VLDL	C	LDL receptor, Apob-100	Turbid or clear
		TG	VLDL overproduction	
			Decreased VLDL degradation	
III	IDL + VLDL	C	ApoE2	Turbid
		TG	VLDL overproduction	
IV[b]	VLDL	C	VLDL overproduction	Turbid or milky
		TG	Decreased VLDL degradation	
V[b]	Chylomicrons	TG	VLDL overproduction	Turbid or milky
	VLDL	C	Decreased VLDL degradation	
			Partial LPL deficiency	
			ApoE4	

[a]TG, Triacylglycerol; C, Total cholesterol.
[b]May also involve a partial deficiency of hepatic lipase.

elevated. Indeed, the term hyperlipoproteinemia denotes only one class of lipid disorders that can more generally be defined as dyslipopoteinemias or dyslipidemias since there may be a deficiency of the particular lipid or lipoprotein rather than an excess.

The cutoff values for defining overt hyperlipidemia (discussed in Chapter 6) can vary in different clinical laboratories and are only approximate guides for physicians. It is now generally accepted that a cholesterol value of over 240 mg/dl for any older age group is above normal; it is recommend that the cholesterol level be below 200 mg/dl.

Except for severe hypertriglyceridemia, which produces early symptoms of abdominal pain, an elevation of plasma lipids rarely causes immediate harmful effects. Rather, the deleterious effects of the hyperlipoproteinemias develop slowly, sometimes requiring many years. The major problem is atherosclerosis, which leads to coronary heart disease and strokes. This raises the question of whether hyperlipidemia can be defined as elevated plasma lipid levels that enhance atherogenesis. However, what these levels are has been difficult to define. Some researchers believe that the correlation between plasma cholesterol levels and atherogenesis is linear over a very wide range of cholesterol levels. Several epidemiological studies suggest that the risk of coronary heart disease begins to increase sharply at cholesterol levels above 200 mg/dl. Thus, hypercholesterolemia can be defined as any level over 200 mg/dl. However, this value is only an approximation, since it does not take into consideration other factors such as age and gender.

An LDL/HDL cholesterol ratio of greater than 3 increases the risk of heart disease. The risk for coronary heart disease increases markedly as the LDL/HDL cholesterol ratio rises above 5.

A link between plasma triacylglycerol level and atherosclerosis has been difficult to determine and there is considerable controversy among researchers. It appears that hypertriglyceridemia is present in a disproportionate number of patients with coronary heart disease. This finding may be attributable in part to the particular lipoprotein species found to be elevated that carries the tri-acylglycerol. Thus, elevation of the triacylglycerol level caused by elevated chylomicrons, may not be as important as if the elevation were due to an eleva-tion of VLDL or IDL or abnormal β-VLDL. An elevation of the VLDL or β-VLDL level will give rise to increased levels of both cholesterol and tri-acylglycerol, but the major culprit is only the cholesterol elevation. Furthermore, one must consider the uptake of these lipoproteins by macrophages and smooth muscle cells and their conversion to foam cells, which are implicated in athe-rogenesis.

5.6. FACTORS LEADING TO HYPERLIPIDEMIAS

Nutrition, genetic makeup, metabolic disorders (related to genetic makeup or nutrition), age, sex, physical activity, and hormonal imbalance can contribute to hyperlipidemia. Some of these factors are interrelated. Epidemiological stud-ies have favored the dietary causation of hyperlipidemia, while clinical investiga-tors and basic scientists have emphasized the genetic factors. Obviously, the dietary factors cannot be dissociated from the genetic factors as is shown below in the discussion of the various types of hyperlipoproteinemias. It is evident that if one overeats, the excess intake of food puts a strain on the metabolic systems of the body even if no genetic abnormality is present and that the strain is much greater if a genetic defect is present. Moreover, the type of food, especially fats, taken in the diet has a profound influence on the metabolic systems that handle the ingested food (discussed in Chapter 7).

5.7. MECHANISMS OF HYPERLIPIDEMIAS

5.7.1. Defects in Lipoprotein Lipase and ApoC-II

Type I hyperlipoproteinemia is associated with an marked elevation of chylomicrons, which produces a severe hypertriglyceridemia, especially when the affected person ingests a high-fat diet. Depending on whether the patient is homozygous or heterozygous for the genetic defect, the plasma triacylglycerol level varies widely, from slightly elevated concentrations of 1000 mg/dl to over 2000 mg/dl. Why these patients do not have elevated VLDL is puzzling but may be due to the fact that nascent VLDL are larger particles that behave like chylomicrons. Persons with type I disease have a genetic defect that leads to a deficiency of LPL or produces a lipase that is inactive or partially active. This causes type Ia disease. Since LPL requires the cofactor apoC-II for activity, a genetic defect in apoC-II produces similar symptoms and gives rise to type Ib disease. Indeed, a new apoC-II variant (apoC-IIBethesda) had been identified (Sprecher et al., 1988). It is of interest that apoC-II was detected in the intestinal mucosa of two patients with apoC-II deficiency (Capruso et al., 1988). This finding shows that the intestine can produce apoC-II but that synthesis of apoC-II in the liver is more important for its function in lipoprotein metabolism.

A major clinical symptom in type I disease is abdominal pain, which results from the restriction of blood flow through capillaries because the very high level of chylomicrons increases the viscosity of the blood. This can cause severe abdominal pain due to swelling of the blood vessels in the omentum of the intestines.

The hypertriglyceridemia in patients with type I disease is improved by restricting dietary fats and replacing the ordinary dietary fats with medium-chain triacylglycerols. These triacylglycerols have fatty acids with chain length of 12 carbon atoms or less. When the medium-chain triacylglycerols are hydrolyzed in the intestines, the fatty acids are absorbed directly into the blood and hence bypass the chylomicron formation. These patients must be supplied with sufficient essential fatty acids (linoleic and linoleic acids) to prevent the development of essential fatty acid deficiency.

5.7.2. Defects in LDL Clearance

Defects in LDL clearance result mainly from genetic defects in the LDL receptor. The research of Goldstein and Brown (1983, 1984; Brown and Goldstein, 1984, 1986, 1987) has elucidated this important lipid abnormality and led to their receiving the Nobel prize in 1985. Genetic defects in the LDL receptor cause type II hyperlipoproteinemia. A deficiency of the LDL receptors was first discovered in patients with FH. Patients with the homozygous form (prevalence of 1 in 1 million) of FH have inherited two defective genes (one from each parent), that code for the mRNA for the LDL receptor. Recent studies have revealed multiple genetic defects that (1) affect synthesis of the receptor and lead to no measurable receptors or very small amounts of receptors, (2) affect transport of the receptor from the ER to the Golgi, (3) affect binding of LDL to the receptor, and (4) affect clustering of the receptors in coated pits. Patients with these genetic defects have type IIa hyperlipoproteinemia evidenced by very high levels of both LDL and plasma cholesterol. Homozygous individuals have cholesterol levels in the range of 800–1200 mg/dl and suffer from early onset of atherosclerosis and coronary heart disease. Cutaneous xanthomas appear within the first 4 years of life, and coronary heart disease can occur at this early age.

Despite the very high level of cholesterol, the fasting plasma of these patients is clear since LDL are small particles and do not scatter light. Patients who are heterozygous (prevalence about 1 in 500) have only one defective gene and hence have half the number of LDL receptors. These patients have cholesterol levels of 400–600 mg/dl. As with the homozygous form, the fasting plasma of these individuals is clear. Many develop atherosclerosis and coronary heart disease later in life, usually in their 50s or 60s but can develop tendinous xanthomas in their 30s.

Some studies have indicated that patients with type II disease may also have an overproduction of LDL, possibly due to an oversynthesis of apoB-100. Research with WHHL rabbits that lack LDL receptors has cast some doubt on this idea. An absence of the apoB/apoE receptor prevents the hepatic uptake of both LDL and IDL so that most of the VLDL is converted to LDL, which can be misconstrued an oversecretion of LDL.

Type IIb hyperlipoproteinemia is associated with increased levels of both plasma cholesterol and triacylglycerols resulting from increased levels of LDL and either IDL or VLDL. This lipid disorder, also called combined hyperlipidemia, is associated with early onset of atherosclerosis and coronary heart disease. Recent work has shown that genetic defects in apoB-100 cause type II hyperlipoproteinemia. This leads to hypercholesterolemia because of elevated levels of LDL, and possibly to some extent, IDL and VLDL, because of the inability of apoB-100 to bind to the LDL receptor.

5.7.3. Defects in ApoE and in Remnant Removal

Recent work has shown the importance of apoEs in the uptake of VLDL remnants (IDL) and in chylomicron remnants. The receptor for chylomicron remnants may recognize apoE whether it is on chylomicron remnants or on VLDL or IDL. As discussed previously, there are three major isoforms of apoE (E2, E3, and E4) that differ in affinity for the LDL receptor. Since there is evidence for a separate apoE receptor that differs from the LDL receptor, these different apoE proteins also may have different affinities for this receptor.

Studies have shown that the affinity of binding of apoEs to the LDL receptor is apoE3>apoE4>apoE2. Since every person inherits two genes for apoE, six phenotypes are possible: E3/3; E3/4; E3/2; E4/4; E4/2; and E2/2. Homozygosity for the phenotype E3/3 occurs in 60% of the human population whereas the phenotype E2/2 is rare and is found only in 1%. Heterozygosity for apoE2 (types E3/2 and E4/2) occurs in about 18% of the population. These apoE isoproteins differ from each other by single amino acid substitutions in their primary structure. In general, the E2 and E4 types are associated with hypertriglyceridemia and hypercholesterolemia, respectively. Indeed, apoE4 is believed to be involved in type V hyperlipoproteinemia (Kuusi et al., 1988). Phenotype E2/2 is associated with type III hyperlipoproteinemia but only 1% of subjects with this phenotype develop hyperlipidemia. Persons with the genotype E2/2 tend to accumulate remnants of chylomicrons and VLDL. However, since the E2/2 pattern rarely causes frank hyperlipidemia, it is a latent defect. Thus, when another defect, such as an increased production of VLDL or an impaired degradation of VLDL (as seen in obesity, diabetes, or hypothyroidism) is present, the latent defect leads to hyperlipoproteinemia type III. In this disease there is an accumulation of VLDL and IDL. The VLDL appear to be abnormal because of the presence of the abnormal apoE. These abnormal VLDL have been called β-VLDL; studies have shown that they are more avidly taken up by macrophages and lead to development of foam cells, which are involved in atherogenesis.

Recent studies by Bradley and Gianturco (1986) have shown that large VLDL from hypertriglyceridemic patients but not VLDL from normal subjects bind to the LDL receptor of human skin fibroblasts because they contain apoE of a correct conformation. These researchers found that apoB-100 was not required for binding of these large VLDL and concluded that apoE is necessary and sufficient for the binding of large VLDL to the LDL receptor.

In a study of the catabolism of chylomicrons in normolipidemic subjects in relation to apoE phenotype, Brenninkmeijer et al. (1987) have found a delayed removal of the both chylomicrons and chylomicron remnants in normolipidemic homozygous E2 persons but not in heterozygous E2 patients. These findings indicate that one normal gene for apoE2 is sufficient for the degradation of chylomicrons and chylomicron remnants. Recent studies on the clearance of chylomicron remnants in normal and hyperlipoproteinemic subjects indicate that the delayed chylomicron remnant clearance in subjects with endogenous hypertriglyceridemia may be largely secondary to overproduction of VLDL particles whose remnants compete with the chylomicron remnants for removal by the liver via apoE receptor-mediated endocytosis.

Wilcox and Heimberg (1987) have studied the secretion and uptake of nascent hepatic VLDL by perfused livers from fed and fasted rats. They reported that during fasting, a reduced secretion of both VLDL and apoE resulted in a VLDL particle that had a lower content of apoE and therefore was taken up by the liver at a lower rate than in fed animals. An increased production of VLDL can be either primary or secondary. The causes of primary overproduction are not well known. Some researchers suggest that excess synthesis of apoB-100 is a primary cause, but this would have to be related to a concomitant overproduction of triacylglycerols, phospholipids, and cholesterol. Secondary lipoprotein overproduction, especially VLDL, is associated with obesity, with diabetes mellitus, and probably with the nephrotic syndrome. High carbohydrate intake and excessive alcohol consumption both stimulate the synthesis of VLDL, but may simply expand the size of the VLDL particles with extra triacylglycerols rather than increase the number of VLDL particles.

Overproduction of apoB-100 coupled to overproduction of triacylglycerols leads to overproduction of VLDL, resulting in endogenous hypertriglyceridemia type IV. This condition is usually accompanied by a decreased clearance of VLDL. Thus, type IV disease results from the combination of overproduction of VLDL and a lipolytic defect in the degradation of VLDL-triacylglycerols. The lipolytic defect can be mild or severe and give rise to a mild or severe increase in VLDL. If the lipolytic defect is severe, it leads to an increase in both chylomicrons and VLDL and thus results in type V hyperlipoproteinemia. Indeed, many patients with type V disease have either diabetes mellitus or obesity. As mentioned above, when overproduction of VLDL is combined with the apoE-2/2 phenotype, the result is type III hyperlipoproteinemia. According to Grundy (1984), some patients have overproduction of VLDL but do not develop hyperlipidemia. They have a high flux rate of VLDL, IDL, and LDL, but because of efficient clearance, the concentrations of these lipoproteins are not increased. Other patients with overproduction of VLDL have defective clearance of either or both IDL and LDL. Kuusi et al. (1988) recently have suggested that apoE4 may be involved in the development of type V hyperlipoproteinemia.

Type IIa disease results mainly from defective receptors for apoB-100 leading to defective clearance of LDL. The more complex type IIb disease is caused by defective receptors for apoB-100 but must also involve a defect in the clearance of VLDL.

The combination of overproduction of VLDL and one or more catabolic defects for the various lipoproteins helps explain how multiple lipoprotein phenotypes can occur within a single family. Families with this condition, which is called familial combined hyperlipidemia, were noted to have a variety of lipoprotein abnormalities (types IIa, IIb, IV, and V).

The field of lipoprotein structure and metabolism is complex and changing rapidly as more specific information is obtained on the structures of the apoproteins and their genetic defects and on the details of how lipoproteins are synthesized and degraded. These new studies will have great impact on the health of humans, since they are closely related to research on atherosclerosis, a condition that predisposes persons to coronary heart disease and cerebral strokes, the primary causes of death in the United States and other highly developed countries.

5.8. LIPOPROTEIN METABOLISM IN DIABETES

The most common form of human diabetes is non-insulin-dependent diabetes (NIDDM), also referred as type II diabetes, maturity onset diabetes, or nonketotic diabetes. It often occurs after age 40 and is commonly associated with obesity. Patients with NIDDM often are hyperinsulinemic but, paradoxically, they are hyperglycemic, which suggests that the insulin is not functioning properly. These patients are said to be insulin resistant. Free fatty acids in plasma are usually elevated in patients with NIDDM.

The most common alteration of plasma lipoproteins in NIDDM is an elevation of VLDL, which leads to an elevation of serum triacylglycerols. Patients who have serum triacylglycerol levels of greater than 300–400 mg/dl usually have other genetic defects in lipoprotein metabolism. The elevation of VLDL has been reported to result from an overproduction or a decreased clearance of VLDL. Overproduction of apoB may also be present, especially in obese patients. An increased influx of free fatty acids and glucose to the liver is believed to lead to an increased production of triacylglycerols and VLDL. Patients with NIDDM therefore have an increased number of VLDL particles that are larger than normal VLDL and are enriched in triacylglycerols. A greater proportion of these "abnormal" VLDL are metabolized without conversion to LDL. A few in vivo studies have shown a decrease in the fractional catabolic rate for LDL in NIDDM due in part to altered binding of the LDL, possibly as a result of nonenzymatic glycosylation of the LDL.

HDL production is decreased in NIDDM in part because of a lower LPL activity accompanied by increased HL activity. The proportion of HDL-2 is decreased, and its composition is altered.

The less common form of human diabetes is insulin-dependent diabetes (IDDM), also called type I diabetes, juvenile onset diabetes, or ketosis-prone diabetes. Onset is usually before age 20 and is characterized by insulin deficiency resulting from an autoimmune destruction of pancreatic B cells. Individuals with IDDM are rarely obese and require insulin therapy, without which they become ketoacidotic. In total insulin deficiency, patients have greatly elevated levels of plasma glucose, ketones, and free fatty acids.

Very high levels of VLDL are common in diabetic ketoacidosis. In untreated IDDM, the fractional catabolic rate for VLDL triacylglycerol is reduced possibly as a result of low levels of lipoprotein lipase coupled to an increased synthesis of triacylglycerols.

A number of studies indicate that plasma HDL levels are low in untreated IDDM. HL activity may be lower even in well-controlled patients, and the HDL2/HDL3 ratio is elevated.

In addition to abnormalities that give rise to elevated levels of certain plasma lipoproteins, there also occur conditions in which there are deficiencies of one or more of these lipoproteins. These abnormalities are called hypolipoproteinemias. There are also abnormalities in which neutral lipids or cholesterol esters accumulate in excess in cells. These abnormalities, along with LCAT deficiency, are discussed below.

5.9. HYPOLIPOPROTEINEMIAS

Hypolipoproteinemias are characterized by a deficiency of one or more apoproteins and lipoproteins. Abetalipoproteinemia is caused by a deficiency of apoB and a lack of chylomicrons, VLDL, and LDL. The clinical manifestations are fat malabsorption, ataxic neuropathy, retinitis pigmentosa, and acanthocytosis. The usual mechanism for transporting triacylglycerols from the intestine and liver to the blood are abolished. All other apoproteins are present in HDL. Inheritance is as an autosomal recessive trait.

In familial hypobetalipoproteinemia, LDL levels are as low as 10–20% of normal because of a decreased synthesis of apoB-100. The trait appears to be autosomal dominant. Recent work (Young et al., 1988) has shown that patients with this disease have a truncated species of apoB. DNA sequencing of genomic clones and enzymatically amplified genomic DNA samples revealed a four-base pair deletion in the apoB gene. This deletion, which results in a frameshift and a premature stop codon, accounts for the truncated apoB species and explains the low apoB and low cholesterol levels in afflicted individuals.

5.9.1. HDL Deficiency (Tangiers Disease)

In Tangiers disease, normal HDL are absent and a small amount of an aberrant type of HDL, designated as HDLT, is present. An imbalanced synthesis of apoAs underlies this rare autosomal recessive disease. The accumulation of cholesterol esters in the reticuloendothelial system accounts for the clinical signs of orange tonsils and hepatosplenomegaly.

5.9.2. LCAT Deficiency

Familial LCAT deficiency (see Chapter 4) is characterized by corneal infiltration, anemia, proteinuria, reduced plasma cholesterol esters and lysolecithin, and increased plasma levels of lecithin and free cholesterol. Blood smears show target cells possibly due to an increased amount of cholesterol in the cell membrane. The disease is believed to be inherited as an autosomal recessive trait.

REFERENCES

Alexander, C. A., Hamilton, R. L., and Havel, R. J., 1976, *J. Cell Biol.*, 69:241.

Armstrong, V. W., Cremer, P., Eberle, E., Manke, A., Schulze, F., Wieland, H., Kreuzer, H., and Seidel, D., 1986, The association between serum Lp(a) concentrations and angiographically assessed coronary atherosclerosis. Dependence on serum LDL levels, *Atherosclerosis,* 62:249.

Aviram,, M., Bierman, E. L., and Oram, J. F., 1989, High density lipoprotein stimulates sterol translocation between intracellular and plasma membrane pools in human monocyte-derived macrophages, *J. Lipid Res.*, 30:65.

Babiak, J., Tamachi, H., Johnson, F. L., Parks, J. S., and Rude, L. L., 1986, Lecithin:cholesterol acyltransferase-induced modifications of liver perfusate discoidal high density lipoproteins from African queen monkeys, *J. Lipid Res.*, 27:1304.

Beaumont, J. L., Carlson, L. A., Cooper, G. R., Fejar, Z., and Fredrickson, D. S., 1970, Classification of hyperlipidaemias and hyperlipoproteinemias, *Bull WHO*, 43:1970

Berg, K., 1963, A new serum type system in man—the Lp system, *Acta Pathol. Microbiol. Scand.*, 59:369.

Bondy, P. K., and Rosenberg, L. E., 1980, *Metabolic Control and Disease*, 8th ed., W. B. Saunders Co., Philadelphia.

Borensztajn, J. (ed.), 1987, *Lipoprotein Lipase*, Evener Publications, Chicago.

Bradley, W. A., and Gianturco, S. H., 1986, ApoE is necessary and sufficient for the binding of large triglyceride-rich lipoproteins to the LDL receptor; ApoB is unnecessary, *J. Lipid Res.*, 27:40.

Brenninkmeijer, B. J., Stuyt, P. M. J., Demacker, P. N. M., Stalenhoef, A. F. H. and van't Laar, A., 1987, Catabolism of chylomicron remnants in normolipidemic subjects in relation to the apoprotein E phenotype, *J. Lipid Res.*, 28:361.

Breslow, J. L., 1987, Genetic regulation of apolipoproteins, *Am. Heart J.*, 113:422.

Brown, M. S., and Goldstein, J. L., 1984, How LDL receptors influence cholesterol and atherosclerosis, *Sci. Am.*, 251:58.

Brown, M. S., and Goldstein, J. L., 1986, A receptor-mediated pathway for cholesterol homeostasis, *Science*, 232:34.

Brown, M. S., and Goldstein, J. L., 1987, Teaching old dogmas new tricks, *Nature*, 330:113.

Brown, M. S., Goldstein, J. L., and Fredrickson, D. S., 1983, Familial type 3 hyperlipoproteinemia, in: *Metabolic Basis of Inherited Disorders*, 5th ed., (J. B. Stanbury, J. B. Wyngaarden, D. S. Fredrickson, J. L. Goldstein, and M. S. Brown, eds.), McGraw-Hill Book Co., New York, Chapter 32, pp. 655-671.

Capurso, A., Morgavero, A. M., Resta, F., Di Tommaso, M., Taverniti, P., Turturro, F., La Rosa, M., Marcovina, S., and Catapano, A. L., 1988, Apolipoprotein C-II deficiency: detection of immunoreactive apolipoprotein C-II in the intestinal mucosa of two patients, *J. Lipid Res.*, 29:703.

Champe, P. C., and Harvey, R. A., 1987, *Lippincott's Illustrated Reviews: Biochemistry*, J. B. Lippincott and Co., Philadelphia.

Chan, B. L., Lisanti, M. P., Rodriguez, E., and Saltiel, A. R., 1988, Insulin-stimulated release of lipoprotein lipase by metabolism of its phosphatidylinositol anchor, *Science*, 241:1670.

Chen, S., Habib, G., Yang, C., Gu, Z., Lee, B. R., Weng, S., Silberman, S. R., Cai, S., Deslypere, J. P., Rosseneu, M., Gotto, A. M. Jr., Li, W., and Chan, L., 1987, Apolipoprotein B-48 is the product of a messenger RNA with an organ-specific in-frame stop codon, *Science*, 238:363.

Cortner, J. A., Coates, P. M., Le, N., Cryer, D. R., Ragni, M. C., Faulkner, A., and Langer, T., 1987, Kinetics of chylomicron remnant clearance in normal and in hyperlipoproteinemic subjects, *J. Lipid Res.*, 28:195.

Curtiss, L. K., Black, A. S., Takagi, Y., and Plow, E. F., 1987, New mechanism for foam cell generation in atherosclerotic lesions, *J. Clin. Invest.*, 86:367.

Eisenberg, S., 1984, High density lipoprotein metabolism, *J. Lipid Res.*, 25:1017.

Gerrity, R. G., 1981, The role of the monocyte in atherogenesis. I. Transition of blood-borne monocytes into foam cells in fatty lesions, *Am. J. Pathol.*, 103:181.

Ghiselli, G., Schaefer, E. J., Gascon, P., and Brewer, H. B., Jr., 1981, Type III hyperlipoproteinemia with apolipoprotein E deficiency, *Science*, 214:1239.

Goldstein, J. L., and Brown, M. S., 1983, Familial hypercholesterolemia, in: *Metabolic Basis of*

Inherited Diseases, 5th ed., (J.B. Stanbury, J.B. Wyngaarden, D.S. Fredrickson, J.L. Goldstein, and M.S. Brown, eds.), McGraw-Hill Book Co., New York, Chapter 33, pp. 672-713.

Goldstein, J. L., and Brown, M. S., 1984, Progress in understanding the LDL receptor and HMG-CoA reductase, two membrane proteins that regulate plasma cholesterol, *J. Lipid Res.*, 25:1450.

Goldstein, J. L., Kita, T., and Brown, M. S., 1983, Defective lipoprotein receptors and atherosclerosis, *N. Engl. J. Med.*, 309:288.

Gotto, A. M., Robertson, A. L., Epstein, S. E., DeBakey, M. D., and McCollum, C. H., 1980, *Atherosclerosis-Diagnosis and Management of Risk Factors for Atherosecerosis*, (H.L. Gross, ed.), The Upjohn Co., Kalamazoo, Michigan, Section I, p. 11.

Gries, A., Fievet, C., Marcovina, S., Nimpf, J., Wurm, H., Mezdour, H., Fruchart, J.C., and Kostner, G. M., 1988, Interaction of LDL, Lp(a), and reduced Lp(a) with monoclonal antibodies against apoB, *J. Lipid Res.*, 29:1.

Grundy, S.M., 1984, Pathogenesis of hyperlipoproteinemia, *J. Lipid Res.*, 25:1611.

Heart-liver transplantation in a child with homozygous familial hypercholesterolemia, 1985, *Nutr. Rev.*, 43:274.

Hobbie, L., Kingsley, D. M., Kozarsky, K. F., Jackman, R. W., and Krieger, M., 1987, Restoration of LDL receptor activity in mutant cells by intercellular junctional communication, *Science*, 235:69.

Hoeg, J. M., Gregg, R. E., and Brewer, H. B., 1986, An approach to the management of hyperlipoproteinemia, *J. Am. Med. Assoc.*, 255:512.

Howard, B. V., 1987, Lipoprotein metabolism in diabetes mellitus, *J. Lipid Res.*, 28:613.

Jonas, A., 1986, Synthetic substrates of lecithin:cholesterol acyltransferase, *J. Lipid Res.*, 276:689.

Jonasson, L., Bondjers, G., and Hansson, G. K., 1987, Lipoprotein lipase in atherosclerosis: its presence in smooth muscle cells and absence from macrophages, *J. Lipid Res.*, 28:437.

Kirchhausen, T., Scarmato, P., Harrison, D. C., Monroe, J. J., Chow, E. P., Mattaliano, R. J., Ramachandran, K. L., Smart, J. E. Ahn, A. H. and Brosius, J., 1987, Clathrin light chains LCA and LCB are similar, polymorphic, and share repeated heptad motifs, *Science*, 236: 320.

Kleinman, Y., Schonfeld, G., Gavish, D., Oschry, Y., and Eisenberg, S., 1987, Hypolipidemic therapy modulates expression of apolipoprotein B epitopes on low density lipoproteins, studies in mild to moderate hypertriglyceridemic patients, *J. Lipid Res.*, 28:540.

Knott, T. J., Rall, S. C. Jr., Innerarity, T. L., Jacobson, S. F., Urdea, M. S., Levy-Wilson, B., Powell, L. M., Pease, R. J., Eddy, R., Nakai, H., Byers, M., Priestly, L. M., Robertson, E., Rall, L. B., Betsholtz, C., Shows, T. B., Mahley, R. W., and Scott, J., 1985, Human apolipoprotein B: structure of carboxyl-terminal domains, sites of gene expression and chromosomal location, *Science*, 230:37.

Knott, T. J., Pease, R. J., Powell, L. M., Wallis, S. C., Rall, S. C. Jr., Innerarity, T. L., Blackhart, B., Taylor, W. H., Marcel, Y., Milne, R., Johnson, D., Fuller, M., Lusis, A. J., McCarthy, B. J. Mahley, R. W., Levy-Wilson, B., and Scott, J., 1986, Complete protein sequence and identification of structural domains of human apolipoprotein B, *Nature*, 323:734.

Kostner, G. M., 1980, Lipoproteine und atherosclerose: einflub der nahrung, *Laboratoriumsblatter*, 30:133.

Kostner, G. M., Avogaro, P., Cazzolato, G., Marth, E., Bittolo-Bon, G., and Quinci, G. B., 1981, Lipoprotein(a) and the risk for myocardial infarction, *Arteriosclerosis*, 38:51.

Kuusi, T., Taskinen, M. R., Solakivi, T., and Makelin, R. K., 1988, Role of apolipoproteins E and C in type V hyperlipoproteinemia, *J. Lipid Res.*, 29:293.

Levy, E., Marcel, Y., Deckelbaum, R. J., Milne, R., Lepage, G., Seidman, E., Bendayan, M., and Roy, C. C., 1987, Intestinal apoB synthesis, lipids, and lipoproteins in chylomicron retention disease, *J. Lipid Res.*, 28:1263.

Li, W., Tanimura, M., Luo, C., Datta, S., and Chan, L., 1988, The apolipoprotein multigene family: biosynthesis, structure-function relationships and evolution, *J. Lipid Res.*, 29:245.

Lusis, A. J., 1988, Genetic factors affecting blood lipoproteins: the candidate gene approach, *J. Lipid Res.*, 29:397.

Mahley, R. W., 1988, Apolipoprotein E: cholesterol transport protein with expanding role in cell biology, *Science*, 240:622.

Mahley, R. W., Innerarity, T. L., Rall, S. C. Jr., and Weisgraber, K. H., 1984, Plasma lipoproteins: apoprotein structure and function, *J. Lipid Res.*, 25:1277.

McLean, J. W., Tomlinson, J. E., Kuang, W. J., Eaton, D. L., Chen, E. Y., Fless, G. M., Scanu, A. M., and Lawn, R. M., 1987, cDNA sequence of human apolipoprotein[a] is homologous to plasminogen, *Nature*, 330:132.

Moore, M. S., Mahaffey, D. T., Brodsky, F. M., and Anderson, R. G. W., Assembly of clathrin-coated pits onto purified plasma membranes, *Science*, 236:558.

Morton, R. E., West, G. A., and Hoff, H. F., 1986, A low density lipoprotein-sized particle isolated from human atherosclerotic lesions is internalized by macrophages via a nonscavenger receptor mechanism, *J. Lipid Res.*, 27:1124.

Murray, R. K., Granner, D. K., Mayes, P. A., and Rodwell, V. W., 1988, *Harper's Biochemistry*, 21st edition, Appleton & Lange, Norwalk, Connecticut.

Naito, H. K., 1988, Apolipoproteins as biochemical markers of cardiac risk, *Am. Clin. Lab.*, January 1988, table 2, p. 28.

Ordovas, J. M., Litwack-Klein, L., Wilson, P. W. F., Schaefer, M. M., and Schaefer, E. J., 1987, Apolipoprotein E isoform phenotypic methodology and population frequency with identification of ApoE1 and ApoE5 isoforms, *J. Lipid Res.*, 28:371.

Powell, L. M., Wallis, S. C., Pease, R. J., Edwards, Y. H., Knott, T. J., and Scott, J., 1987, A novel form of tissue-specific RNA processing produces apolipoprotein B-48 in intestine, *Cell*, 50:831.

Rapacz, J., Hasler-Rapacz, J., Taylor, K. M., Checovich, W. J. and Attie, A. D., 1986, Lipoprotein mutations in pigs are associated with elevated cholesterol and atherosclerosis, *Science*, 234:1573.

Schneider, W. J., Beisiegel, U., Goldstein, J. L., and Brown, M. S., 1982, Purification of the low density lipoprotein receptor, an acidic glycoprotein of 164,000 molecular weight, *J. Biol. Chem.*, 257:2664.

Schneider, W. J., 1989, The low density lipoprotein receptor, *Biochim. Biophys. Acta*, 988:303 (review article).

Schonfeld, H., and Krul, E. S., 1986, Immunologic approaches to lipoprotein structure, *J. Lipid Res.*, 27:583.

Segrest, J. P., Jackson, R. L., Morrisett, J. D., and Gotto, A. M., 1974, A molecular theory of lipid-protein interactions in the plasma lipoproteins, *FEBS Lett.*, 38:247.

Soria, L. F., Ludwig, E. H., Clarke, H. R. G., Vega, G. L., Grundy, S. M., and McCarthy, B. J., 1989, Association between a specific apolipoprotein B mutation and familial defective apolipoprotein B-100, *Proc. Natl. Acad. Sci. USA*, 86:589.

Sprecher, D. L., Taan, L., Gregg, R. E., Fojo, S. S., Wilson, D. M., Kashyap, M. L., and Brewer, H. B. Jr., 1988, Identification of an apoC-II variant (apoC-IIBethesda) in a kindred with apoC-II deficiency and type I hyperlipoproteinemia, *J. Lipid Res.*, 29:273.

Sudhof, T. C., Goldstein, J. L., Brown, M. S., and Russell, D. W., 1985. The LDL receptor gene: a mosaic of exons shared with different proteins, *Science*, 228:815.

Tall, A. R., 1986, Plasma lipid transfer proteins, *J. Lipid Res.*, 27:361.

Tall, A. R., and Small, D. M., 1978, Plasma high-density lipoproteins, *N. Engl. J. Med.*, 299:1232.

The lipid research clinics coronary primary prevention trial results, reduction in incidence of coronary heart disease, II. The relationship of reduction in incidence of coronary heart disease to cholesterol lowering, 1984, *J. Am. Med. Assoc.*, 251:351.

Tria, E., and Scanu, A. M. (eds.), 1969, *Structural and Functional Aspects of Lipoproteins in Living Systems*, Academic Press, New York.

Utermann, G., 1987, Apolipoprotein E polymorphism in health and disease, *Am. Heart J.*, 113:433.

Utermann, G., 1989, The mysteries of lipoprotein(a), *Science*, 246:904.

Walton, K. W., 1975, Pathogenic mechanisms in atherosclerosis. *Am. J. Cardiol.*, 35:542.

Wilcox, H. G., and Heimberg, M., 1987, Secretion and uptake of nascent very low density lipoprotein by perfused livers from fed and fasted rats, *J. Lipid Res.*, 28:351.

Wion, K. L., Kirchgessner, T. D., Lusis, A. J., Schotz, M. C., and Lawn, R. M., 1987, Human lipoprotein lipase complementary DNA sequence, *Science*, 235:1638.

Yamamoto, T., Bishop, R. W., Brown, M. S., Goldstein, J. L., and Russell, D. W., 1986, Deletion of cysteine-rich region of LDL receptor impedes transport to cell surface in WHHL rabbit, *Science*, 232:1230.

Yang, C., Chen, S., Gianturco, S. H., Bradley, W. A., Sparrow, J. T., Tanimura, M., Li, W., Sparrow, D. A., DeLoof, H., Rosseneu, M., Lee, F., Gu, Z., Gotto, A. M., Jr., and Chan, L., 1986, Sequence, structure, receptor-binding domains, and internal repeats of human apolipoprotein B-100, *Nature*, 323:738.

Young, S. G., Northey, S. T., and McCarthy, B. J. 1988, Low plasma cholesterol levels caused by a short deletion in the apolipoprotein B gene, *Science*, 241:591.

Zannis, V. I., Breslow, J. L., Utermann, G., Mahley, R. W., Weisgraber, K. H., Havel, R. J., Goldstein, J. L., Brown, M. S., Schonfeld, G., Hazzard, W. R., and Blum, C., 1982, Proposed nomenclature of ApoE isoproteins, ApoE genotypes, and phenotypes, *J. Lipid Res.*, 23:911.

Zilversmit, D. B., 1973, A proposal linking atherogenesis to the interaction of endothelial lipoprotein lipase with triglyceride-rich lipoproteins, *Circul.*, 33:633.

Chapter 6

CHOLESTEROL, LIPOPROTEINS, AND ATHEROSCLEROSIS

6.1. ATHEROSCLEROSIS

Atherosclerosis, a major health problem in the United States, is the accumulation of lipid-rich plaques in the inner lining (intima) of arteries of the body. Advanced plaques are the result of an accumulation of cholesterol, cholesterol esters, phospholipids, live and dead cells, calcium, and other components including collagen, elastin, and proteoglycans. Plaques can be large enough to severely restrict the flow of blood to tissues or can narrow the lumen of arteries to a degree that enhances formation of a thrombus that plugs the artery and leads to a coronary heart attack or to stroke.

Heart diseases rank number 1 and cerebrovascular diseases rank number 3 with respect to the number of deaths per 100,000 population (Table 6-1). These major diseases are related to atherosclerosis, abnormal lipid metabolism, and platelet aggregation.

Traits that may predict the occurrence of coronary heart disease (CHD) are called risk factors. A risk factor is a measure that is correlated with the risk that a given individual will develop a disease. Risk factors are usually identified in prospective studies in a large population by observing the correlation between the putative risk factor and the occurrence of the disease. These factors are not necessarily direct causes of the disease nor does their presence guarantee that disease will occur in a given person. Moreover, their absence does not indicate that the disease will not occur. The probability of developing CHD increases with the number and severity of the positive risk factors in a given individual. The three primary risk factors for CHD and stroke are high blood pressure (systolic pressure 140 mm Hg and diastolic pressure 90 mm Hg), elevated blood cholesterol (especially a high LDL-cholesterol and a low HDL-cholesterol), and smoking. (Table 6-2). Overweight (obesity), diabetes, sex, age, and personality

Table 6-1. Causes of Death in the United States in 1984[a]

Cause of death	Deaths per 100,000 (rank)
All causes (total)	866.7
Unintentional injuries	40.1 (4)
Malignant neoplasms	191.6 (2)
Heart diseases	324.4 (1)
Suicide, homicide	20.6 (7)
Congenital anomalies	5.6 (10)
Prematurity	3.5 (11)
Sudden infant death syndrome	2.4 (12)
Cerebrovascular diseases	65.6 (3)
Chronic liver diseases and cirrhosis	11.3 (9)
Pneumonia and influenza	25.0 (6)
Chronic obstructive pulmonary diseases	29.8 (5)
Diabetes mellitus	15.6 (8)

[a]From the Centers for Disease Control, 1986.

type are secondary risk factors. It is noteworthy, however, that CHD can occur without known risk factors.

Persons with Friedman type A personality appear to be at higher risk for heart disease. Type A personality is characterized by impatience, urge to meet deadlines, hostility toward others, and tendency to become readily angered or excited by trivial matters. Although elevated levels of hormones such as epinephrine, testosterone, and ACTH may be links between type A personality and CHD, these links are not well defined. Platelet aggregation and high blood pressure, which are exacerbated by elevated levels of epinephrine, may be one

Table 6-2. Risk Factors for CHD

Modifiable	Fixed
Hypercholesterolemia	Genetic mutations
Hypertension	Age
Smoking	Sex (male)
Diabetes mellitus	Family history of CHD
Low HDL-cholesterol (<35 mg/dl)	
High LDL-cholesterol (>160 mg/dl)	
Obesity (30% overweight)	
Personality type A	

link. Whether type A personality is a risk factor for CHD is still highly controversial.

Atherosclerosis affects millions of humans. The disease can begin early in life, is usually a very slow process, and predisposes one to developing CHD. In 1986 about 200,000 coronary artery bypass operations were done in the United States. This figure rose to about 250,000 in 1988. This operation costs from $20,000–$35,000 per person and thus represents several billion dollars of health expense to the nation.

An understanding of the process of atherogenesis is vital to effective treatment of heart disease and stroke. This process is complex, involving abnormal metabolism of cholesterol and lipoproteins, platelet aggregation, hormones, prostaglandins, hemodynamic stress, and other factors. The relationship between elevated plasma cholesterol levels and the risk of atherosclerosis and CHD has been substantiated by a number of studies on laboratory animals, epidemiological studies between and within human populations, genetic studies, and clinical trials.

6.1.1. Early History of Atherosclerosis

Aortas of Egyptian mummies showed the presence of atherosclerotic plaques. The plaques were called atheroma from the Greek athere (mush). In 1818, Chevruel identified cholesterol in human bile and named it from the Greek chole (bile) and steros (solid). Vogel identified cholesterol in plaques in 1843. In the 1950s, Berchow, Duguid, French, Packman, and Mustard helped develop the response to injury hypothesis to explain plaque formation. In 1900, Windaus found that atheromas contained 7 times as much cholesterol and 25 times as much cholesterol ester as did normal arteries. The thrombogenic theory of atherogenesis was proposed by Rokitansky in 1952, when the role of platelets in thrombus formation became known.

6.1.2. Experimental Atherosclerosis

Animals develop atherosclerosis when fed diets that raise plasma cholesterol levels. In 1913 the Russians Ignatowski and Anitschkow fed rabbits diets very high in cholesterol. The rabbits developed atheromas very similar to those found in humans. Since then, numerous studies involving feeding of cholesterol-rich diets have been carried out in chickens, dogs, and monkeys. All of these studies showed that high-cholesterol diets induced atherogenesis.

A major question is whether atherosclerosis is a reversible process. It now appears quite clear from both animal and human studies that atherosclerotic lesions undergo slow regression when plasma cholesterol levels are reduced below 190 mg/dl in humans by combined therapy with drugs, diet, and exercise, and by cessation of smoking.

6.1.3. Epidemiological Data

Population studies have shown that persons on low-fat diets have less CHD than do those on high-fat diets, especially diets high in saturated fats and cholesterol. South African Bantus eat very low fat diets, have a mean plasma cholesterol of 160 mg/dl, and have a very low incidence of CHD. Yemenite Jews and Japanese who migrate to the United States initially have a lower incidence of CHD but eventually after adopting American diets high in fat, develop CHD at the same rate as do Americans.

The lipid content of human aortic intima increases with age. In particular cholesterol esters derived mainly from LDL accumulate in the intima of arteries with age. Studies have shown that damaged arterial intima take up LDL more rapidly than do normal intima due in large part to the infiltration of the intima with medial smooth muscle cells and monocytes.

A large body of epidemiologic data, including comparisons between various populations throughout the world, support a direct relationship between plasma cholesterol levels and the rate of development of atherosclerosis and CHD. It is noteworthy, however, that premature CHD can result from high plasma cholesterol levels even in the absence of other risk factors. This is very clearly demonstrated in children who have the rare homozygous familial hyper-cholesterolemia. These children have mutated receptors that do not allow LDL in the blood to be taken up by cells. This defect leads to very high levels of LDL (and hence cholesterol) in the blood, since LDL are the cholesterol-rich lipoproteins. Cholesterol levels in the plasma of these children are 1000–1300 mg/dl rather than in the normal range of 140–190 mg/dl. These children develop atherosclerosis and severe CHD very early in life (by age 4–6) and can be helped only by liver and heart transplantation.

6.2. SURVEY STUDIES: CHOLESTEROL AND CHD

Several large surveys have revealed a positive correlation between level of plasma cholesterol and risk of CHD. This correlation is attributable to the atherogenic effect of elevated plasma levels of cholesterol as illustrated in prospective autopsy studies showing a linear correlation between the concentration of plasma cholesterol and the severity of atherosclerosis (Fig. 6-1).

Several surveys, such as the Framingham Heart Study, the Pooling Project, and the Israeli prospective study, all confirm that plasma cholesterol level is correlated significantly with the prevalence of CHD. The combined results of

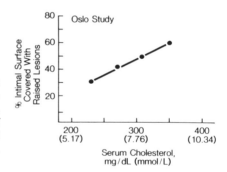

Figure 6-1. Relationship between premortem plasma cholesterol level and severity of atherosclerosis at autopsy in prospective study. Atherosclerosis is expressed as percentage of surface of coronary arteries covered with raised atherosclerotic lesions. (From Grundy, 1986. Reproduced with permission.)

these studies are shown in Fig. 6-2. The rates for CHD are relatively constant for cholesterol levels up to 200 mg/dl but above this threshold the risk for CHD increases as the concentration of cholesterol rises. Some interpret these data to mean that little is gained by way of protection against CHD by lowering the cholesterol level below 200 mg/dl. However, a much larger study, the Multiple Risk Factor Intervention Trial (MRFIT), did not support the concept of a strict threshold.

The MRFIT involved 356,222 men who were 35–57 years old. They were followed for 6 years and the number of deaths caused by CHD was correlated with their plasma cholesterol level. The relationship was positive throughout and curvilinear (Fig. 6-3). At higher levels of plasma cholesterol (250 mg/dl) mortality increased more rapidly. The magnitude of the increased risk was fourfold in the top 10% as compared with the bottom 10%.

A difficulty in demonstrating a correlation between total plasma cholesterol and risk of CHD with values below the mean for any population relates to the opposing effects on risk of cholesterol carried by LDL and HDL. The risk of CHD increases as the level of LDL-cholesterol increases (especially to >150 mg/dl), but the risk falls as the level of HDL-cholesterol increases (especially to

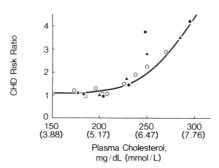

Figure 6-2. Relationship between plasma cholesterol level and relative risk of CHD in three prospective studies: Framingham heart study (solid circles); Pooling Project (triangles); and Israeli prospective study (open circles). (From Grundy, 1986. Reproduced with permission.)

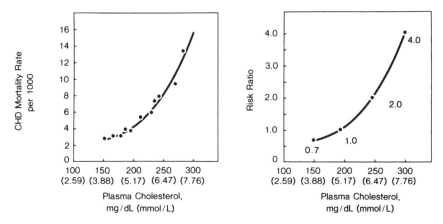

Figure 6-3. Relationship between plasma cholesterol concentration and coronary mortality in MRFIT participants. (A) Coronary mortality for all individuals who were normotensive at screening expressed as yearly rates per 1000; (B) Coronary mortality expressed by risk ratios. (From Grundy, 1986. Reproduced with permission.)

>40 mg/dl). The total cholesterol level in plasma is a good first approximation for assessing risk of CHD. However, the risk is better defined by examining the individual lipoprotein fractions, especially LDL-cholesterol and HDL-cholesterol. Moreover, in persons with type III hyperlipidemia, it is important to assess the level of β-VLDL-cholesterol. In the fasting state, with a plasma that after standing overnight in the refrigerator is turbid but has no upper creamy layer (i.e., has no chylomicrons), the VLDL-cholesterol level is approximately one-sixth the plasma triacylglycerol level.

The role of HDL in protecting persons from premature CHD is receiving considerable attention today. The first report from the Framingham study that demonstrated an inverse relationship between HDL-cholesterol and incidence of CHD was based on 4 years of surveillance. The subjects, ages 49–82, were followed for 12 years. Participants at the 80th percentile of HDL-cholesterol were found to have half the risk of developing CHD when compared with subjects at the 20th percentile of HDL-cholesterol (Castelli et al., 1986). Indeed, subjects who had HDL-cholesterol levels above 60 mg/dl had less risk of developing CHD even when their total cholesterol levels varied from less than 200mg/dl to 260 mg/dl. There are several provisional concepts as to how HDL may exert its protective effect. HDL, in particular HDL2, may retard the development of a plaque by inhibiting the uptake of LDL by endothelial cells of arteries. Another idea involves the role of HDL in reverse cholesterol transport. In this mechanism, HDL, via its interaction with the enzyme LCAT, transfers excess cholesterol from nonhepatic cells directly to the liver or to VLDL, IDL, and LDL, which are then taken up by the liver via specific receptor-mediated endocytosis; the excess cholesterol is subsequently converted to bile acids. Since HDL contain different species of apoA (apoA-I, -II, -III, and -IV), the question arises which of these apoproteins is important for the antiatherogenic effect of HDL.

6.3. ALCOHOL INTAKE AND CHD

Modest intake of alcohol (one or two drinks of wine per day) has been shown to increase the level of plasma HDL. The association between alcohol intake and risk of CHD is U shaped with both nondrinkers and heavy drinkers having a higher incidence of CHD than moderate users of alcohol (see Chapters 3 and 7). The mechanism whereby alcohol intake increases plasma HDL and offers some protection against CHD is not understood. A recent study has shown that moderate intake of alcohol increases the synthesis of apoA-I in liver. Since apoA-I is a major protein of HDL, this finding explains in part the rise in plasma HDL in persons who drink modest amounts of alcohol.

Alcohol influences membrane fluidity. Part of the action of alcohol may be mediated by a fluidity change that increases the functioning of the LDL receptors, which increases the uptake of plasma LDL. Alcohol may also influence the structure of the lipoproteins, which in turn may affect their interaction with membrane receptors that mediate lipoprotein uptake by cells. Another action of HDL is to facilitate the transfer of apoC-II to chylomicrons and VLDL thereby allowing LPL to degrade these lipoproteins.

Alcohol intake leads to an increase in plasma triglycerides, primarily as VLDL. Elevated levels of plasma triglycerides are believed to be an independent risk factor for CHD (Castelli, 1986), especially in women and in men with HDL levels below 40 mg/dl.

CHD is a disease of multifactorial etiology, and there are a number of other risk factors to consider when treating a patient (Table 6-2). It is obvious that some of the modifiable factors have a genetic basis. For example, persons with type IIa hypercholesterolemia have genetic defects of the LDL receptors and those with type III hyperlipidemia have genetic defects of the apoE proteins.

The relationship between smoking, high cholesterol levels, and CHD is illustrated by data from the MRFIT study. In Fig. 6-4, mortality rates for CHD are plotted against plasma cholesterol levels for normotensive subjects and for smokers and nonsmokers in the normotensive category. Although the risk ratios at different cholesterol levels are similar for each curve, the absolute differences in risk are much higher for smokers.

Because of the positive correlation between plasma cholesterol level and risk of CHD, many investigators believe that the average cholesterol level for the whole population should be as low as possible. The question of what constitutes

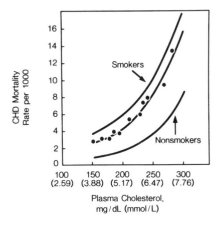

Figure 6-4. Yearly coronary mortality rates (per 1000) vs plasma cholesterol level for normotensive smokers and nonsmokers among MRFIT participants. The intermediate curve is that for the whole population. (From Grundy, 1986. Reproduced with permission.)

an ideal plasma cholesterol level was addressed by a group of epidemiologists, clinical investigators, and experimental pathologists. They proposed that the ideal cholesterol levels for adults be set at 130–190 mg/dl. For the sake of simplification and reality, the ideal cholesterol level was recommended not to exceed 190 mg/dl. Table 6-3 shows the new categories for total plasma cholesterol levels and the levels of LDL-cholesterol in adults as they relate to risk of CHD.

High-risk patients should begin a program of dietary therapy that includes restriction of foods high in cholesterol and saturated fats, substituting saturated fats with polyunsaturated fats, restricting total dietary fats to no more than 30% of total daily calories, and restricting total daily calories to prevent overweight. Lean meats such as veal, chicken breast, turkey breast, and fish should replace fatty meats such as hamburgers, sausage, and fatty steaks. If diet does not reduce

Table 6-3. Total Plasma Cholesterol and LDL-Cholesterol Levels in Adults

Category	Level (mg/dl)
Total plasma cholesterol	
Desirable	<200
Borderline high	200–239 (no other risk factors)
High	>240 mg/dl
LDL-cholesterol	
Desirable	<130
Borderline high	130–159
High	>160

Table 6-4. How Dietary Lipids Influence Plasma Cholesterol Levels

Factor	Effect
Dietary cholesterol	Suppresses the synthesis of LD receptors, increases cholesterol level
Dietary saturated fat	Reduces activity and level of LDL receptors, increases synthesis of VLDL, increases cholesterol and triacylglycerol levels
Obesity	Increases synthesis of VLDL, increases cholesterol and triacylglycerol levels

plasma cholesterol levels to the safe range, then drug therapy should be considered. The manner in which diet influences plasma cholesterol levels is shown in Table 6-4. It is noteworthy that dietary cholesterol and saturated fat decrease the level of LDL receptors and elevate blood cholesterol level.

6.4. DRUG THERAPY FOR HYPERLIPIDEMIAS

When dietary attempts to reduce plasma cholesterol level are not successful, appropriate drug therapy is initiated. Three factors determine when to use drug therapy: failure to respond adequately to dietary therapy; LDL-cholesterol level of >190 mg/dl; and LDL-cholesterol level of >160 mg/dl in men if one other risk factor is present; in women if two other risk factors are present.

Treatment of the hyperlipidemias with lipid-lowering drugs and antithrombic and fibrinolytic drugs is discussed in Chapter 8.

6.5. POSTULATED MECHANISMS OF ATHEROSCLEROSIS

The average adult human heart beats about 70 times per minute and pumps approximately 4000 gallons of blood per day through about 60,000 miles of blood vessels. It is estimated that the daily work load of the heart can lift a man to the top of the Empire State Building. The constant pumping of this large volume of blood imposes stresses and wear and tear on the various blood cells and on the endothelial cells of arteries. The relentless flow of blood through the vessels can damage platelets, erythrocytes, white blood cells, and endothelial cells and set the stage for the slow development of atheroma (lipid plaques). This situation is aggravated in persons who have high levels of plasma LDL, high blood pressure, and platelets that are more prone to aggregation. The development of atheroma and atherosclerosis represents a very complex process that proceeds over many years and involves the interplay of plasma lipoproteins, prostaglandins, certain types of white blood cells, platelets, and hemodynamic factors.

The intima of the artery is composed of endothelial cells that act as a barrier to the blood proteins and lipoproteins. The epithelium also produces vaso-relaxing substances, epithelium-derived relaxing factor (EDRF), and pros-tacyclins which counteract thromboxane, ADP, and serotonin, the vaso-constricting substances produced by platelets. Beneath the intima lies a sheet of elastic fibers called the internal elastic lamina that separates the intima from the media. The media consists of smooth muscle cells, collagen, elastic fibers, and proteoglycans. The adventicia, or the outermost layer of the artery, consists of fibroblasts, smooth muscle cells, and capillaries arranged between bundles of collagen and proteoglycans. It is separated from the media by the external elastic lamina.

Lipids in plaques and cells give the fatty streak a yellow color. By age 10, most humans have some fatty streaks. Between the ages of 10 and 25, the area of the intima covered by fatty streaks increases from about 10% to about 30–50%. The whitish fibrous plaque is elevated and consists of smooth muscle cells and foam cells. The cells are surrounded by collagen, elastic fibers, and proteogly-cans. The cells and the matrix of collagen and proteoglycans form the fibrous cap. The complicated lesion appears to be a fibrous cap that has undergone calcification, hemorrhage, necrosis, and mural thrombosis. The complicated lesion severely narrows the lumen of the artery which impedes blood flow, enhances the formation of a thrombus, and can lead to a heart attack or stroke.

The postulated events leading to atherosclerosis and heart disease or stroke are depicted in Table 6-5. Damage to the intima, leading to development of an atheroma, has been called the response to injury hypothesis. Damage to the intima is believed to be caused by several factors, including high blood pressure, which increases vortex flow and high lateral pressure in the artery, by high levels of plasma LDL, by smoking, and by autoimmune reactions. How smoking and high LDL produce the damage is not known. Some believe that smoking creates carbon monoxide and other chemicals that cause anoxia to the endothelial cells. High LDL levels and possibly high levels of β-VLDL may lead to increased uptake by endothelial cells, which causes cell damage. Recent work has indi-cated that atherosclerotic plaques contain lipoprotein particles called A-LDL. These LDL particles are more electronegative, have a lower hydrated density, and have a lower protein/lipid ratio than do plasma LDL. Moreover, the apoB in the A-LDL is highly degraded. The A-LDL particles are recognized and taken up by macrophages via a nonreceptor scavenger mechanism. Damage to endothelial cells can be produced by chemical agents such as homocystine and by physical agents such as a balloon catheter. These latter agents have been used experimen-tally in animals to induce atherosclerosis.

The involvement of platelets in atherogenesis is a crucial point of the throm-

Table 6-5. Postulated Events Leading to Atherosclerosis
and Heart Disease or Stroke[a]

The intima is damaged by chemical or mechanical factors
Platelets bind to the subendothelium
Binding leads to platelets producing thromboxane A_2 and ADP
Thromboxane A_2 and ADP stimulate platelet aggregation
Aggregated platelets release PDGF and inflammatory agents
Aggregated platelets form a thrombus plug
PDGF stimulates migration and growth of medial smooth muscle cells
Smooth muscle cells migrate to the intima, ingest LDL, and β-VLDL, and proliferate
Monocytes invade the inflamed area and ingest LDL and β-VLDL
Monocytes and smooth muscle cells become filled with cholesterol and other lipids
Monocytes and smooth muscle cells become foam cells that initiate plaque formation
Smooth muscle cells secrete collagen, elastin, and proteoglycans, which enhance plaque formation
Plaques grow into complicated lesions containing lipids, dead cells and calcium
Plaque growth narrows the artery or occludes the artery and a thrombus plug occurs at the narrowed artery causing a heart attack or stroke.

[a]These events are provisional but are based on a considerable body of scientific data.

bogenic hypothesis. Platelet aggregation leads to the release of several factors including platelet-derived growth factor (PDGF), serotonin, epinephrine, thromboxane A_2, ADP, and calcium. Platelet aggregation is enhanced by platelet binding to collagen or other macromolecules such as proteoglycans that have been exposed at the site of endothelial injury. Epinephrine, ADP, and thromboxane A_2 act to further enhance platelet aggregation. PDGF stimulates medial smooth muscle cells to multiply and migrate to the site of injury on the intima, where they take up LDL particles and become foam cells. Smooth muscle cells also synthesize and secrete collagen, elastin, and proteoglycans, which form the matrix of the fibrous plaque. This matrix may further enhance the binding of platelets, leading to platelet aggregation which results in either plaque growth or a thrombus (see Chapter 8).

It is noteworthy that many persons can have a normal life span even if they have moderate or severe atherosclerosis. The crucial event is when atherosclerosis leads to the formation of a thrombus that occludes a blood vessel and causes a stroke or heart attack. Whereas atherosclerosis is a slow process usually developing over many years, a thrombus can form within seconds or minutes. Thrombus formation involves platelets and other clotting factors, and prevention requires the use of specific drugs separate from those used to treat hyperlipidemia.

LPL has been found in the fibrous cap of atherosclerotic plaques, where more than half of the cells have the enzyme. Most of the LPL is localized in smooth muscle cells. In contrast, macrophages present in the plaque have very little lipoprotein lipase. Zilversmit (1983) has proposed a hypothesis implying an atherogenic role for LPL. The hypothesis states that some of the cholesterol-rich remnants formed by the action of LPL may be taken up in the intima of arteries. The fact that endothelial cells do not produce LPL but smooth muscle cells do suggests that smooth muscle cells play an important role in forming cholesterol deposits from the cholesterol-rich remnants formed by the action of LPL on chylomicrons and VLDL.

Platelets have the ability to induce cholesterol ester accumulation in cultured human monocyte-derived macrophages. Thrombin-activated platelets or substances released by activated platelets increase both the rate of esterification of cholesterol and the accumulation of cholesterol esters in these cells. A consequence of an increase in the cholesterol ester content of macrophages is a dramatic increase in their synthesis and secretion of apoE. Takagi et al. (1988) have recently found that platelet-induced secretion of apoE paralleled the capacity of platelets to induce macrophage cholesterol accumulation. This finding provides new insight into the possible role of platelet-enhanced apoE production in atherosclerosis.

However, Desai et al. (1989) have reported that the occupation of cell-surface receptors that bind apoE-rich HDL (HDL-E) particles impairs platelet responsiveness to exogenous agonists and that platelet aggregation in the presence of whole HDL may reflect the relative concentrations of the individual subclasses of HDL, some that enhance and others that inhibit platelet aggregation. It remains to be discovered how the other apoE-containing lipoproteins behave toward platelets.

It is obvious that the process of atherosclerosis is complex and is only partially understood. Many of the current concepts on the mechanisms of atherogenesis are provisional and undoubtedly will be modified. The postulated role of Lp(a) in atherogenesis is discussed in Chapter 5.

REFERENCES

Anderson, K. M., Castelli, W. P., and Levy, D., 1987, Cholesterol and mortality: 30 years of follow-up from the Framingham study, *J. Am. Med. Assoc.*, 257:2176.

Atmeh, R. F., Stewart, J. M., Boag, D. E., Packard, C. J., Lorimerm, A. R., and Shepherd, J., 1983, The hypolipidemic action of probucol: a study of its effects on high and low density lipoproteins, *J. Lipid Res.*, 24;588.

Benditt, E. P., and Schwarz, S. M., 1984, Atherosclerosis: what can we learn from studies in human tissues, *Lab. Invest.*, 50:3.

Blank, D. W., Hoeg, J. M., Kroll, M. H., and Ruddel, M. E., 1986, The method of determination must be considered in interpreting blood cholesterol levels, *J. Am. Med. Assoc.*, 256:2867.

Brown, M. S., and Goldstein, J. L., 1984, How LDL receptors influence cholesterol and atherosclerosis, *Sci. Amer.*, 251:58.

Brown, M. S., and Goldstein, J. J., 1986, A receptor-mediated pathway for cholesterol homeostasis, *Science*, 232:34.

Castelli, W. P., 1986, The triglyceride issue: a view from Framingham, *Am. Heart J.*, 112:432.

Castelli, W. P., Garrison, R. J., Wilson, P. W. F., Abbott, R. D., Kalousdian, S., and Kannel, W. B., 1986, Incidence of coronary heart disease and lipoprotein cholesterol levels: the Framingham study, *J. Am. Med. Assoc.*, 256:2835.

Centers for Disease Control, 1986, Causes of death in the United States in 1984, *Morbid, Mortal, Weekly Report*, 35:457.

Coronary heart disease without risk factors, 1989, *Nutr. Rev.*, 47:18.

Desai, K., Bruckdorfer, K. R., Hutton, R. A., and Owen, J. S., 1989, Binding of apoE-rich high density lipoprotein particles by saturable sites on human blood platelets inhibits agonist-induced platelet aggregation, *J. Lipid Res.*, 30:831.

Etingin, O. R., Weksler, B. B., and Hajjar, D. P., 1986, Cholesterol metabolism is altered by hydrolytic metabolites of prostacyclin in arterial smooth muscle cells, *J. Lipid Res.*, 27:530.

Expert panel, 1988, Report of the national cholesterol education program expert panel on detection, evaluation, and treatment of high blood cholesterol in adults, *Arch. Intern. Med.*, 148:36.

Goldbourt, U., Holtzman, E., and Neufeld, H. N., 1985, Total and high density lipoprotein cholesterol in the serum and risk of mortality: evidence of a threshold effect, *Brit. Med. J.*, 290:1239.

Gotto, A. M., Robertson, A. L., Epstein, S. E., DeBakey, M. D., and McCollum, C. H., 1980, in: *Atherosclerosis* (H.L. Gross, ed.), pp. 1—54, The Upjohn Co., Kalamazoo, Michigan.

Grundy, S.M., 1986, Cholesterol and coronary heart disease, *J. Am. Med. Assoc.*, 256:2849.

Grundy, S. M., 1986, Cholesterol and coronary heart disease: a new era, *J. Am. Med. Assoc.*, 256:2849.

Harlan, J. M., and Haker, L. A., 1983, *Thrombosis and Coronary Artery Disease*, The Upjohn Co., Kalamazoo, Michigan.

Hegele, R. A., Huang, L., Herbert, P. N., Blum, C. B., Buring, J. E., Hennekens, C. H., and Breslow, J. L., 1986, Apolipoprotein B-gene DNA polymorphisms associated with myocardial infarction, *N. Engl. J. Med.*, 315:1509.

Jonasson, L., Bondjers, G., and Hansson, G. K., 1987, Lipoprotein lipase in atherosclerosis: its presence in smooth muscle cells and absence from macrophages, *J. Lipid Res.*, 28:437.

Kannel, W. B., 1983, High-density lipoproteins: epidemiologic profile and risks of coronary artery disease, *Am. J. Cardiol.*, 52:9B

Kromhout, D., Bosscheiter, E. B., and De Lezenne Coulander, C., 1985, The inverse relation between fish consumption and 20-year mortality from coronary heart disease, *N. Engl. J. Med.*, 312:1205.

Kuo, P. T, Kostis, J. B., Moreyra, A. E., and Hayes, J. A., 1981, Familial type II hyperlipoproteinemia with coronary heart disease. Effect of diet-colestipol-nicotinic acid treatment, *Chest*, 79:286.

Lipoproteins and coronary heart disease, 1984, *Clin. Nutr.*, 3:131.

Martin, M. J., Browner, W. S., Hulley, S. B., Kuller, L. H., and Wentworth, D., 1986, Serum cholesterol, blood pressure, and mortality: implications from a cohort of 361,622 men, *Lancet*, ii:933.

Molecular biology of coronary heart disease, 1987, *Nutr. Rev.*, 45:108.

Multiple risk factor intervention trial research group: multiple risk factor intervention trial. Risk factor changes and mortality results, 1982, *J. Am. Med. Assoc.*, 248:1465.

Pooling project research group. Relationship of blood pressure, serum cholesterol, relative weight, and ECG abnormalities to incidence of major coronary events: final report of the Pooling Project, 1978, *J. Chron. Dis.*, 31:201.

Olson, R. E., 1986, Mass intervention vs. screening and selective intervention for the prevention of coronary heart disease, *J. Am. Med. Assoc.*, 255:2204.

Rapp, J. H., Connor, W. E., Lin, D. S., Inahara, T., and Porter, J. M., 1983, Lipids of human atherosclerotic plaques and xanthomas: clues to the mechanism of plaque formation, *J. Lipid Res.*, 24:1329.

Roberts, A. B., Lees, A. M., Strauss, H. W., Fallon, J. T., Taveras, J., and Kopiwoda, S., 1983, Selective accumulation of low density lipoproteins on damaged arterial wall, *J. Lipid Res.*, 24:1160.

Ross, R., and Glomset, J. A., 1976, The pathogenesis of atherosclerosis, *N. Engl. J. Med.*, 295:369, 420.

Rudel, L. L., Parks, J. S., Johnson, F. L., and Babiak, J., 1986, Low density lipoproteins in atherosclerosis, *J. Lipid Res.*, 27: 465.

Scanu, A. M., Wissler, R. W., and Getz, G. S. (eds.), 1979, *The Biochemistry of Atherosclerosis*, Marcel Dekker, New York.

Siegel, D., Grady, D., Browner, W. S., and Hulley, S. B., 1988, Risk factor modification after myocardial infarction, *Ann. Intern. Med.*, 108:213.

Stamler, J., Wentworth, D., and Neaton, J. D., 1986, MRFIT Research Group, is relationship between serum cholesterol and risk of premature death from coronary heart disease continuous and graded? Findings in 356,222 primary screenes of the MRFIT, *J. Am. Med. Assoc.*, 256:2823.

Starzi, T.E., Bilheimer, D. W., Bahnson, H. T., Shaw, B. W. Jr., Hardesty, R. L., Griffith, B. P., Iwatsuki, S., Zitelli, B. J., Gartner, J. C. Jr., Malatack, J. J., and Urbach, A. H., 1984, Heart-liver transplantation in a patient with familial hypercholesterolemia, *Lancet*, i:1382.

Strong, J. P., 1986, Coronary atherosclerosis in soldiers. A clue to the natural history of atherosclerosis in the young, *J. Am. Med. Assoc.*, 256:2863.

Takagi, Y., Dyer, C. A., and Curtiss, L. K., 1988, Platelet-enhanced apolipoprotein E production by human macrophages: a possible role in atherosclerosis, *J. Lipid Res.*, 29:859.

The lipid research clinics coronary primary prevention trial results. I.Reduction in incidence of coronary heart disease, 1984, *J. Am. Med. Assoc.*, 251:351.

Utermann, G., 1989, The mysteries of lipoprotein(a), *Science*, 246:904.

Wynder, E. L., Field, F., and Haley, N. J., 1986, Population screening for cholesterol determination, a pilot study, *J. Am. Med. Assoc.*, 256:2839.

Yamamoto, A., Matsuzawa, Y., Kishino, B., Hayashi, R., Hirobe, K., and Kikkawa, T., 1983, Effects of probucol on homozygous cases of familial hypercholesterolemia, *Atherosclerosis.*, 48:157.

Chapter 7

DIETARY MANAGEMENT OF ELEVATED BLOOD LIPIDS

7.1. DIETARY LIPIDS, HYPERLIPIDEMIA, AND HEART DISEASE

High levels of plasma cholesterol, in particular LDL-cholesterol, are associated with premature atherosclerosis and heart disease. Therefore, it is important to know how drug therapy and dietary management can be used to lower blood cholesterol levels. The effect of dietary management, particularly of dietary lipids, will be considered first, since for most people this approach to treatment of of hyperlipidemia is tried before drug therapy.

In the United States, the public is being informed by the American Heart Association and other agencies of the effect of nutrition on heart disease. This effort has led to very significant changes in the eating habits of Americans. Americans are consuming less eggs, beef, pork, hamburgers, and hotdogs and more chicken, turkey, and fish. The recommended dietary goals for Americans with regard to the contributions of protein, carbohydrate, and fat to daily caloric intake, are shown in Table 7-1.

The general goal of dietary therapy is to reduce elevated levels of plasma cholesterol and triacylglycerols while maintaining a nutritionally adequate eating pattern and maintaining a desirable weight. The step-one diet (Table 7-1) is tried first; if it is not effective, the step-two diet is initiated. These dietary regimens are tried before drug therapy is contemplated.

Polyunsaturated fatty acids include linoleic, linolenic, and arachidonic acids, which occur primarily in vegetable oils, and eicosapentenoic and docosahexenoic acid, which occur primarily in fish oils. The major monounsaturated fatty acid is oleic acid, which is found in high amounts in olive oil. Restriction of dietary intake of saturated fatty acids is important in reducing plasma cholesterol levels. Saturated fatty acids occur in high amounts in dairy products, bacon fat, and lard; in hydrogenated fats such as Crisco and Spry; and in certain vegetable oils such as coconut oil and palm oil. More detailed information on recom-

Table 7-1. Dietary Management of Hyperlipidemia[a]

Nutrient	Recommended intake (% of total calories)	
	Step-one diet	Step-two diet
Total fat	<30	<30
Saturated FA[b]	<10	<7
Polyunsaturated FA	Up to 10	Up to 10
Monounsaturated FA	10 to 15	10 to 15
Carbohydrates	50 to 60	50 to 60
Protein	10 to 20	10 to 20
Cholesterol	<300 mg/d	<200 mg/d

[a]From National Cholesterol Education Program Expert Panel, 1988.
[b]FA, fatty acid.

mended dietary modifications to lower blood cholesterol can be found in the article by the Expert Panel of the National Cholesterol Education Program (1988).

The degree of reduction of LDL-cholesterol levels that can be achieved by dietary therapy depends on a person's eating habits before the diet is started and on the inherent responsiveness of the person. In general, people with high cholesterol levels show a greater absolute reduction in total and LDL-cholesterol levels than do those with relatively low cholesterol levels. Metabolic ward studies suggest that switching from the typical American diet to the step-one diet could reduce cholesterol levels on average by 30–40 mg/dl. Advancing to the step-two diet can be expected to cause a further reduction of about 15 mg/dl.

Dietary fiber also is important for lowering blood cholesterol. Fiber, which consists of complex carbohydrate polymers containing mucilages, pectins, lignin, and hemicelluloses, is found in plant foods such as grains, vegetables, legumes, fruits, nuts, and seeds. Fiber can be divided into two types, soluble and insoluble. Insoluble fiber is coarse and chewy and is considered roughage. Soluble fiber is sticky like gums and gels. Fiber is not digested by humans, since humans lack the enzymes that degrade fiber. Insoluble fiber regulates bowel regularity by absorbing water and swelling. This softens the stool and stimulates the intestinal muscles to pass the stool through the intestine. Fibers are believed to lower blood cholesterol by binding bile acids and increasing their passage through the small intestines. Since most of the bile acids are absorbed in the small intestine, the less time food stays in the small intestine, the lower the amount of bile acids will be absorbed. The loss of bile acids then triggers the liver to convert more cholesterol to bile acids. Since the liver utilizes cholesterol in the blood, especially as LDL, as more LDL-cholesterol is taken up by the liver, it is converted to bile acids, which lowers the cholesterol level in the blood.

Soluble fiber appears to be more effective than insoluble fiber in lowering blood cholesterol. Preliminary research indicates that soluble fiber found in corn bran, oat bran, carrots, and apples can reduce blood cholesterol. Usually, persons with higher cholesterol levels appear to have a greater lowering of cholesterol levels than do those with normal cholesterol levels. Reductions of blood cholesterol by 3–20% have been reported. To reduce blood cholesterol by 15–20%, a substantial amount of fiber (three large bowls of oat bran per day) must be consumed. This type of diet is not practical. One cup of oat bran per day may lower blood cholesterol by approximately 3%. It is estimated that most Americans eat 10–15 g of total dietary fiber per day. The National Cancer Institute and the American Diabetes Association encourage 20–30 g per day. To get this latter amount of fiber, a person must consume three to five servings of whole-grain breads and cereals, three servings of vegetables, and two to three servings of fruit per day. A serving is considered to be two slices of bread, one bagel, one cup of rice or pasta, one medium piece of fruit, or one-half to two-thirds cup of vegetables. Foods highest in fiber are black-eyed peas, kidney beans, All Bran, corn bran, oat bran, corn, peas, baked potatoes with skin, raw apples, barley, almonds, graham crackers, and corn bread.

7.1.1. Dietary Lipids

Dietary fats represent a variety of lipid molecules, and each must be considered. A recommendation "to lower dietary fat" has little meaning, since the type of fat being selected is critical. By "dietary fat" does one refer to cholesterol, phospholipids, or triacylglycerols? Dietary triacylglycerols are the major type of dietary fat occurring in animal and vegetable foods and must be distinguished, since they contain different amounts and different types of fatty acids (i.e., saturated and unsaturated) and cholesterol. The fat contents of selected seafood, poultry, and meats are given in Table 7-2. In general, lamb, beef, and pork are higher in fat content than seafood and poultry.

The cholesterol contents of some foods are given in Table 7-3. It is noteworthy that fruits, grains, and vegetables do not contain cholesterol, since cholesterol is uniquely an animal fat. The foods highest in cholesterol are egg yolk, beef liver, and beef kidney.

Animal and vegetable foods that are sources of saturated, monounsaturated, and polyunsaturated fatty acids are presented in Tables 7-4, and percentages of the specific types of polyunsaturated fatty acids are shown in Table 7-5. Safflower, sunflower, corn, and soybean oils have the highest contents of polyunsaturated fatty acids. Rapeseed and olive oil have the highest content of monounsaturatred fatty acid. Palm oil, coconut oil, beef, lard, and butter have the highest

Table 7-2. Fat Content of Selected Seafood, Poultry, and Meats

Source	Percent fat
Seafood	
Cod, haddock, pollock, orange roughy, blue shark, surf clams, scallops, flounder, monkfish, sea bass, soft shell clams, northern shrimp, crabs, lobster, American and spiny striped bass, Pacific halibut, brook trout, ocean perch, mussels, Atlantic halibut, swordfish, pink salmon	5 or less
Rainbow trout, bluefish, coho salmon, sardine, albacore tuna, herring, whitefish mackerel, lake sturgeon, king salmon, eel	5.1–17
Poultry	
Turkey, flesh only, roasted; chicken, broiler/fryer flesh only, roasted	5–7.4
Meat	
Veal, lean cut; lamb, beef, pork	7.9–13.2

content of saturated fatty acids. Safflower, corn, cottonseed, and sunflower have the highest contenst of 18:2 ω-6 fatty acids and seafood is highest in 20:5 ω-3 and 22:6 ω-3 fatty acids. Linseed oil is uniquely high in 18:3 ω-3 fatty acids.

Vegetable fats do not contain cholesterol but rather have a variety of other plant sterols. Animal fats, especially beef, butter, liver, kidney, and sweet-

Table 7-3. Cholesterol Content
of Some Common Foods[a]

Food[b]	Cholesterol (mg)
Fruits, grains, vegetables	None
Scallops	53
Oysters	45
Clams	65
Fish, lean	65
Chicken, turkey, light meat	80
Lobster	85
Beef, lean	90
Chicken, turkey, dark meat	95
Crab	100
Shrimp	150
Egg yolk (one)	275
Beef liver	440
Beef kidney	700

[a]From U.S. Department of Health and Human Services, 1985.
[b]Based on about 3.5 oz of cooked meat.

Table 7-4. Sources of Some Commonly
Used Fats[a]

Vegetable oils high in polyunsaturated fatty acids
  Rapeseed
  Safflower
  Sunflower
  Corn
  Soybean
  Peanut
  Canola
Vegetable oil high in monounsaturated fatty acid
  Olive
Vegetable oils high in saturated fatty acids
  Palm
  Coconut
Animal fats containing cholesterol and high in saturated fatty acids
  Beef
  Lard
  Butter
  Pork

---

[a]Vegetable oils do not contain cholesterol but instead contain other plant sterols such as sitosterol and ergosterol.

breads, contain relatively high amounts of cholesterol. Animal fats (triacyl-glycerols) usually contain less polyunsaturated fatty acids than do vegetable fats. The exceptions are coconut and palm oil, which have relatively high levels of saturated fatty acids. Moreover, fish triacylglycerols contain higher amounts of ω-3 polyunsaturated fatty acids (eicosapentenoic and docosahexenoic acid) than do triacylglycerols of beef, pork, or poultry. These factors are important, since the type of fatty acids taken in the diet determines whether these fatty acids lower, raise, or do not affect cholesterol levels in humans. The amount of cholesterol taken in the diet also influences the level of blood cholesterol but primarily only when the dietary intake of cholesterol exceeds 300 mg/day.

7.1.2. Effects of Dietary Fatty Acids on Plasma Lipids and Lipoproteins

Substitution of monounsaturated and polyunsaturated fats (in proper amounts) for saturated fats in the diet has a lipid-lowering effect in most humans. Monounsaturated fatty acids, particularly oleic acid, have been shown to (1) reduce plasma LDL levels when substituted for saturated fatty acids, (2) reduce the risk of CHD, (3) not suppress the immune system or promote carcinogenesis in animals, (4) not raise plasma triacylglycerol levels, and (5) not lower plasma HDL levels. (These findings were reported at the Second Colloquium on

Table 7-5. Dietary Sources of Polyunsaturated Fatty Acids[a]

Source	18:2 ω-3	18:3 ω-3	20:4 ω-6	20:5 ω-3	22:6 ω-3
Mainly ω-6					
Safflower	73	0.5			
Corn	57	1.0			
Cottonseed	50	0.4			
Sunflower	56	0.3			
Peanut	29	1.0			
Mainly ω-3					
Linseed	15	55			
Salmon	1	1		8	5
Cod liver	2	1		12	12
Catfish	6	0.7	2	4	9
Mackerel	2	1	2	10	16
Whale	1		4	3	7
ω-6 and ω-3					
Soybean	51	7			
Walnut	55	11			
Low in ω-6 and ω-3					
Cow milk fat	2	1			
Human milk fat	7	0.7	0.2	0.6	0.3
Lard	10	1			
Chicken fat	17	1			
Beef tallow	4	0.5			
Egg yolk	11	0.2	6		
Beef liver	10	0.5	6		
Coconut oil	2				
Olive oil	8	0.7			
Cocoa butter	3	0.2			
Palm oil	9	0.3			

[a]From Goodnight et al., 1982. Reproduced with permission.

Monounsaturates-Role of Monounsaturated Fatty Acids in Human Nutrition, held on February 26, 1987, in Bethesda, Maryland.) However, the monounsaturated fatty acids cannot be converted to prostaglandins and leukotrienes, as can the polyunsaturated fatty acids. Another potential benefit of monounsaturated fatty acids is their inhibition of lipid peroxidation (Balasubramanian et al., 1988; Diplock et al., 1988). Lipid peroxidation is a major cause of damage to cell membranes and is believed to enhance carcinogenesis.

The polyunsaturated C18 ω-6 fatty acids (linoleic and arachidonic acid), the C18 ω-3 fatty acid (linolenic acid), the C20 ω-3 fatty acid (eicosapentenoic acid), and the C22 ω-3 fatty acid (docosahexenoic acid) lower blood levels of cholesterol. Of particular interest is the finding that only the ω-3 fatty acids from fish oils lower triacylglycerol levels both in normal persons and in those with elevated levels of triacylglycerols (in the form of VLDL and/or chylomicrons).

On the other hand the ω-6 fatty acids may increase blood levels of triacylglycerols in some individuals. Dietary ω-3 fatty acids from fish have a strong
triacylglycerol-lowering effect in both type 1 and type 2 hyperlipoproteinemia.
The mechanisms of these lipid-lowering effects not fully understood, but some
studies indicate they are related to increased loss of bile acids and cholesterol
from the body and to inhibition of fatty acid synthesis in the liver. The unsaturated fatty acids also may enhance the activity or increase the level of the LDL
receptors. Dietary saturated triacylglycerols rich in palmitic and stearic acid tend
to increase blood lipid levels. It is clear that to speak only in broad terms of
dietary fat or unsaturated fatty acids as a means of regulating blood lipid levels is
no longer sufficient; one must specify the type and amount of unsaturated fatty
acid and its dietary source.

7.1.3. Effects of Dietary Fatty Acids on Platelet Aggregation

The point just mentioned becomes more important when one considers the
other major effect of certain polyunsaturated fatty acids as antithrombic agents.
While feeding of both ω-6 and ω-3 fatty acids leads to an inhibition of platelet
aggregation in laboratory tests, only fish oils having ω-3 fatty acids give a
prolonged bleeding time in humans as judged by the Ivy test, in which the
bleeding time of a small cut in the skin is measured. Why is the antithrombic
effect of fish oils important? The answer relates to the role of platelets in forming
a thrombus, or clot. Two major factors, atherosclerosis and thrombus formation,
lead to occlusion of arteries and subsequent heart attack or stroke. An artery may
be occluded by 50–60% without causing a major problem since most tissues
have an excess of blood flow. However, if a vessel that is 50–60% occluded is
then subjected to thrombus formation, the thrombus will immediately shut off all
blood flow and lead to a heart attack or stroke. Therefore, the prolonged intake of
ω-3 fatty acids not only lowers blood cholesterol levels which protects from
atherosclerosis, but also inhibits thrombus formation. Thus, these fish fatty acids
exert a doubly protective effect against heart attack and stroke.

7.1.4. Role of Dietary ω-3 Fatty Acids in Heart Disease

Polyunsaturated fatty acids, which contain three or more double bonds of
which the double bond farthest from the carboxyl (COOH) end is three carbon
atoms from the terminal methyl end of the molecule, are called ω-3 fatty acids.
The important dietary ω-3 fatty acids are those containing 18, 20, and 22 carbon
atoms. An 18-chain fatty acid is linolenic acid, which is designated as 18:3 ω-3.
This fatty acid occurs in high amounts in fish oils and in linseed oil. Other
vegetable oils, such as corn, safflower, sunflower, olive, and cottonseed oils,
contain very little 18:3 ω-3 fatty acid but are high in 18:2 ω-6 (linoleic) fatty
acid. Both linoleic acid and linolenic acid are essential dietary constituents for

humans. The human body can convert linoleic acid to arachidonic acid (20:4 ω-6) and can convert linolenic acid to eicosapentenoic acid (20:5 ω-3) and to docosahexenoic acid (22:6 ω-3). Therefore, the essential fatty acids taken in the diet can be converted to longer-chain polyunsaturated fatty acids by enzymes that elongate the carbon chain and other enzymes that introduce double bonds. It must be realized, however, that these enzyme reactions are limited with regard to how fast they work and the extent to which they can synthesize the longer-chain fatty acids. Therefore, linolenic acid taken in the diet may not be converted in sufficient amount to eicosapentenoic acid and docosapentenoic acid. Consumption of fish will provide these latter two fatty acids.

Fish oils are the primary source of the ω-3 fatty acids (20:5 ω-3 and 22:6 ω-3), whereas the vegetable oils (except linseed oil) and animal fats (beef, pork, and poultry) are the primary source of the ω-6 fatty acids (18:2 ω-6 and 20:4 ω-6). The importance of this difference in fatty acid composition becomes apparent when one considers the conversion of polyunsaturated fatty acids to prostaglandins.

7.1.5. Conversion of Polyunsaturated Fatty Acids to Prostaglandins

Many cells of the body contain enzymes that convert polyunsaturated fatty acids to prostaglandins. The prostaglandins were so named because these biologically powerful agents were first isolated from the prostate gland. Figure 7-1 shows the conversion of polyunsaturated fatty acids to prostaglandins and leukotrienes. The structures of prostaglandins and leukotrienes are given in Chapter 1.

It is noteworthy that only dietary fats from fish contain high amounts of the C20 and C22 ω-3 fatty acids (eicosapentenoic and docosahexenoic acids), whereas these acids are very low in vegetable oils and animal fats such as beef, pork and poultry. Of the vegetable oils, only linseed oil has a high content of the C18 ω-3 fatty acid (linolenic acid). However, linseed oil is not very palatable and is not a common dietary source of lipid.

The importance of the dietary ω-3 fatty acids is understood when one considers the biological functions of the prostaglandins with respect to platelet aggregation. The prostaglandins represent several types of molecules in which the polyunsaturated fatty acid has been modified enzymatically by cyclization, hydroxylation, and ketolization.

The most extensively studied prostaglandins are types 2 and 3. For our purposes, we shall consider only the prostacyclins and thromboxanes since they play an important role in platelet aggregation. The sites of synthesis and effect of these prostaglandins on platelet aggregation are given in Table 7-6.

The table shows that whereas both prostacyclin I_2 (PGI_2) and PGI_3 inhibit platelet aggregation, only thromboxane A_2 (TXA_2) stimulates platelet aggregation. Since the prostacyclin/thromboxane ratio appears to be critical in regulating platelet aggregation, it is apparent that the PGI_3/TXA_3 ratio will be more favor-

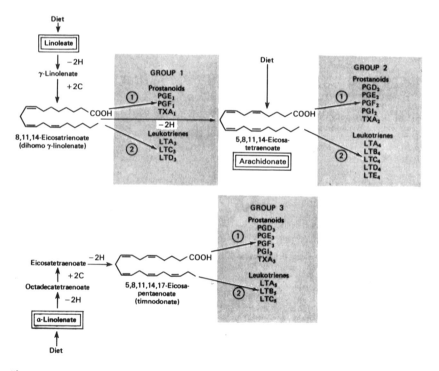

Figure 7-1. Conversion of dietary polyunsaturated fatty acids to prostaglandins and leukotrienes. PG, Prostaglandins; PGI, prostacyclin; TX, thromboxane; LT, leukotriene; 1, cyclooxygenase pathway; 2, lipoxygenase pathway. (From Murray et al., 1988. Reproduced with permission.)

able than the PGI_2/TXA_2 ratio, since TXA_3 has no effect on aggregation. Therefore, the beneficial effect of dietary fish in protecting humans from thrombosis and heart attack is due to the fact that fish contain high amounts of eicosapentenoic acid, which provides a favorable PGI_3/TXA_3 ratio. Platelet aggregation is believed to be involved in the genesis of atheroma by releasing PDGF that causes

**Table 7-6. Sites of Synthesis and Effects
of Prostaglandins on Platelet Aggregation**

Prostaglandin	Site of synthesis	Effect on platelet aggregation
TXA_2	Platelet	Stimulates
PGI_2	Endothelial cell	Inhibits
TXA_3	Platelet	None
PGI_3	Endothelial cell	Inhibits

migration and proliferation of medial smooth muscle cells into the intima (see Chapter 6).

Not only do fish oils protect by inhibiting platelet aggregation, but they also have a hypolipidemic effect and lower plasma LDL and VLDL levels (see Chapter 5). This effect provides another mechanism for preventing or minimizing plaque formation. Recent work also has shown that eicosapentenoic acid in fish oil reduces the production of platelet-activating factor (PAF) in platelets and leukotriene-B_4 in leukocytes. PAF stimulates platelet aggregation and secretion of vasoactive amines, whereas leukotriene-B_4 induces inflammatory response in leukocytes.

The low death rate from CHD among Greenland Eskimos may be explained in part by their high consumption of fish (an average of about 400 g/day). In Japan, the lowest death rate from coronary artery disease occurs in Okinawa, where fish consumption is about twice as high as it is on the mainland.

A recent study in the Netherlands (Nutr. Rev., 1984) showed that mortality from CHD was more than 50% lower among men who consumed at least 30 g of fish per day than among men who did not eat fish. A random sample of 1088 men was selected in 1960. All were born between 1900 and 1919 and had lived in the area of Zutphen for at least 5 years. Of the 1088 men, 872 took part in a medical examination and in a dietary survey. A validation study of the diet history survey was carried out on 49 men whose diets were weighed and chemically analyzed. Serum cholesterol, blood pressure, smoking habits, anthropometric measurements, physical activity, and occupation were determined. The vital status of all 1088 men enrolled in the study was verified after 20 years of follow-up. During the 20 years, 390 men died, 132 from cancer, 110 from CHD, and 33 from cerebrovascular disease. Of the 852 men who were free from coronary artery disease at the time of entry into the study, 78 died from CHD during the 20-year period.

Each participant was placed in one of five categories based on daily fish consumption: 0, 1–14, 15–29, 30–44, and 45 or more g. For these categories, risk ratios and 95% confidence limits were calculated. A chi-square value for trend was calculated to determine whether there was a dose–response relationship between fish consumption and death from coronary artery disease.

In 1960, the average fish consumption of Zutphen men was 20 g/day (range, 0–90 g/day). About two-thirds of the fish consisted of lean fish (cod and plaice) and one-third consisted of fatty fish (herring and mackerel). No relationship was found between fish consumption and major risk factors such as age, plasma cholesterol, blood pressure, and cigarette smoking. The average total serum cholesterol in 1960 was 230 mg/dl with the 10th percentile value 170 mg/dl and the 90th percentile value 290 mg/dl. Fish consumption was unrelated to subcapsular skinfold thickness, physical activity, occupation, or energy intake. Significant relationships were found in the intake of some nutrients.

Of the variables recorded, only age, fish consumption, and serum cholesterol were significantly related to death from CHD during the 20-year period. Fish consumption varied inversely to incidence of CHD after adjustment for other risk factors. The risk ratios for death from CHD decreased with increasing fish consumption and were about 2.5 times lower among men who consumed more than 30 g of fish per day. These results suggest that eating one or two fish meals per week may be of value in minimizing CHD.

Platelet aggregation has been shown to be lower in Japanese fishermen than in farmers and lower in Eskimos than in Danes. Reduced platelet aggregation and longer bleeding times have been found in humans fed increased amounts of eicosapentenoic acid (2–20 g/day). In the Netherlands study, the 90 g of fish per day would contain about 0.5 g of eicosapentenoic acid.

More studies are required to pinpoint the components of fish oils that give this beneficial effect and to see how consumption of fish oils compares with the use of antithrombogenic agents such as aspirin and indomethacin. (The effects of these antithrombic agents are discussed in Chapter 8.)

The dietary regimen for helping people with elevated blood cholesterol or triacylglycerol levels should include substitution of fish and white meat of chicken or turkey for beef, pork, and other fatty meats and restriction of foods that are high in cholesterol (e.g., egg yolk, ice cream, liver, sweetbreads, and cheese). Foods rich in saturated fat, such as pie crust, cakes, and coconut, and hydrogenated vegetable oils (Crisco and Spry), coconut oil, and palm oil should also be restricted. The diet should also substitute certain vegetable oils rich in the polyunsaturated linoleic or linolenic acids in place of butter, cream, and hydrogenated fats. Whole milk and cream should be replaced by low-fat (1 or 2% fat) or skim milk.

The nutritional value of margarines as a replacement for butter is not fully established, since the atherogenic effects of modified (*trans*) polyunsaturated fatty acids in margarine (as a result of hydrogenation) are in conflict. However, low-fat margarines contain fewer calories (on a weight basis) than butter and margarines are higher in unsaturated fatty acids.

A secondary and unrelated benefit of a low-fat (less than 25% of total daily calories) diet is that it may afford some protection against certain types of cancer (colon, stomach, and breast cancer). These findings are based mainly on studies in animals, but it seems reasonable to extrapolate them to humans. Thus, it is prudent to restrict the total fat intake even when replacing saturated fat with polyunsaturated fat. Total fat intake should not exceed 30% of total calories. This limitation is particularly important for people who have types 2b, 3, 4, or 5 hyperlipoproteinemia since their bodies cannot handle diets high in fat. People with type 1 disease must have a diet severely restricted in fat, but care must be taken to prevent a deficiency of the essential fatty acids (EFA). Medium-chain triglycerides are a useful dietary supplement for people with type 1 hyperlipoproteinemia.

The possible beneficial effects of increasing the amount of polyunsaturated fats in the diet to help lower blood cholesterol levels and minimize heart attacks must be weighed against any possible increased risk of cancer. Dietary therapeutic measures must be followed in a reasonable, moderate manner that avoids drastic changes. It appears best to consume no more than the minimum amount of polyunsaturated fats required to protect a person at risk against CHD; intake of greater amounts might negate the beneficial effect by increasing the risk of cancer. Moreover, the risk of cancer may be minimized by following a low-fat diet. Why do diets high in fat increase the risk of cancer? One answer is that fats act as procarcinogens. That is, they enhance the carcinogenic effects of certain drugs that cause cancer. The cancer-inducing drugs of major concern are the fat-soluble drugs. Another hypothesis is that very high fat diets, especially those high in polyunsaturated fatty acids, may increase membrane fluidity which makes cells more susceptible to cancer-causing agents. Much more research is needed to clarify these important problems in nutrition. Do rancid fats and peroxidized fats produced by cooking increase the risk of colon cancer? Which specific fats are procarcinogenic? Do *trans* fatty acids produced during hydrogenation of vegetable oils have harmful effects on the body? Do antioxidant vitamins such as vitamin E and vitamin A protect humans from cancer? The answers to these questions remain to be determined.

7.1.6. Relation of Dietary Lipids to CHD

In villages in Japan and Yugoslavia where the mean total cholesterol level is 160 mg/dl, the incidence of heart attacks was found to be less than 5/1,000 men over a 10-year period. In eastern Finland where the mean total cholesterol level is 265 mg/dl, the incidence of heart attacks is 14 times higher! In populations with intermediate cholesterol levels, the incidence of heart attacks is between these two extremes. Populations consuming low amounts of animal fats, such as meats and dairy products, tend to have lower cholesterol levels (lower LDL) than do populations consuming high amounts of animal fats. Japanese have lower cholesterol levels than Americans and also have less heart disease. However, when Japanese move to the United States and adopt an American diet, their rate of heart disease becomes the same as that of Americans. Many other studies in humans and animals indicate that elevated levels of cholesterol are directly related to heart disease. Studies in humans are based on general population statistics, which include many variables that often make their statistical significance difficult to assess.

The LDL receptor system helps explain in part the effects of diet on cholesterol levels. A high intake of animal fat is associated with a high intake of cholesterol; this intake allows cholesterol to accumulate in the liver, leading to decreased synthesis of LDL receptors. Diets high in saturated animal fats increase the levels of plasma cholesterol by increasing the synthesis of VLDL and reducing the level of LDL receptors in the liver. When the receptor levels are decreased, the body cannot degrade LDL very effectively; as a result LDL levels rise.

The hyperlipidemia in type 3 and 4 disease is quite responsive to diets low in total calories and low in saturated fat and cholesterol. Recent animal studies have shown that low- fat diets are associated with high levels of LDL receptors as a result of increased synthesis of these receptors. When animals (dogs and rabbits) are fed diets high in cholesterol, their synthesis of LDL receptors is markedly reduced, leading to an elevation of both IDL and LDL. These animals become more susceptible to atherosclerosis.

Recent studies in humans also have shown that the number of LDL receptors decreases with age. The reason for this is not entirely known but is related to hormonal changes, dietary changes, and decreased physical activity. Estradiol and thyroxine are hormones known to stimulate the synthesis of LDL receptors in the liver. The reduction of these hormones with age may help explain the reduction of LDL receptors and concomitant increases in plasma cholesterol with age. Cholesterol levels in women tend to rise after menopause, in large part because of a decrease in estrogen production.

Studies indicate that cultured cells bind LDL optimally at LDL concentrations below 50 mg/dl. However, in the United States and other industrialized countries, the average normal level of LDL in adults is about 125 mg/dl. Therefore, one might ask whether dietary therapy alone can reduce LDL levels sufficiently to be of substantial help to humans. This approach can be effective but it requires a substantial restriction of saturated fats and cholesterol in fatty meats, eggs, and other dairy products, a very difficult solution. For many persons whose bodies can handle these fatty food, the dietary restriction will be very modest. For others, however, especially those genetically predisposed to hyperlipoproteinemia, the dietary restriction will be severe, and a combined regimen of drugs, dietary therapy, and exercise should be tailored to each individual.

7.1.7. Influence of Egg Consumption on Blood Cholesterol

Egg yolk has a high cholesterol content (Table 7-4). The effect of dietary cholesterol on serum cholesterol levels has been studied quite extensively. No association was seen between cholesterol intake and serum cholesterol level when the intake varied from 200 to 1500 mg/day. However, it must be emphasized that some of these statistical correlations do not take into account individual variations. For example, a person with type 2, 3 or 4 hyperlipidemia may be very susceptible to cholesterol intake, whereas normal people are not. The type of diet and the baseline level of serum cholesterol are important factors to consider. The dietary factors to be reckoned with are the amount and type of other lipids, especially polyunsaturated fatty acids and saturated fatty acids, and the amount of vegetables containing plant sterols, which influence cholesterol absorption. Some researchers believe that the absorption of cholesterol is regulated in the intestine. This hypothesis suggests that the body has a sensing mechanism such that as the blood cholesterol level rises, less cholesterol is absorbed from the diet.

Dietary cholesterol influences cholesterol synthesis in the liver. Cholesterol synthesis in the liver is regulated at the level of the rate-limiting step catalyzed by the enzyme HMGR, which converts HMG-CoA to mevalonic acid. As the level of cholesterol rises as a result of dietary intake, there is a negative feedback control that reduces the activity of HMGR.

The effects of egg yolk feeding in three males and three females has been investigated (Nutr. Rev., 1985). The subjects were 26–42 years old, weighed 123–178 lb, and had total serum cholesterol levels of 136–215 mg/dl (mean, 191 ± 27 mg/dl). They were asked to consume normal diets but to omit eggs, butter, shellfish, and organ meats and to limit intake of meat and fish to 100 g/day for 10 days. Six eggs per day were then added to this basal diet for 10 more days. Energy intake increased from 2260 ± 524 to 2689 ± 96 kcal/day and dietary cholesterol increased from 207 ± 26 to 1803 ± 155 mg/day. Dietary fat increased from 39 ± 5% of total calories to 46 ± 6%. The polyunsaturated fatty acid/saturated fatty acid (P/S) ratio remained constant at 0.47 ± 0.6.

In both trials the average total serum cholesterol decreased from 191 ± 27 to 168 ± 21 mg/dl in the initial 10-day period. Eating six eggs per day increased total serum cholesterol by 13%. LDL cholesterol increased by 18 and 24% and HDL2 levels increased by 23 and 49% in the first and second trials.

The effect of dietary cholesterol and fatty acids on plasma lipoprotein were studied in 20 young men over a period of 4–6 weeks (Nutr. Rev., 1985). In all subjects total serum cholesterol and LDL levels were lower on a basal diet containing 300 mg cholesterol per day than on an ad libitum diet. Adding three eggs/day (750 cholesterol) to a diet with a P/S ratio of 0.4 raised LDL cholesterol by 16±14 mg/dl. Six eggs per day (1500 mg cholesterol) increased LDL-cholesterol by 17±22 mg/dl. However, on a diet with a P/S ratio of 2.5, neither the three or six eggs/day produced significant effects on serum cholesterol. These studies indicate that diets high in cholesterol or with a low P/S ratio increase the level of LDL-cholesterol in serum.

Feeding one extra large egg and one placebo per day to 17 lactovegetarians (4 men, 13 women) for 3 weeks was studied in a randomized , double-blind crossover experiment (Nutr. Rev., 1985). Adding one egg per day increased dietary cholesterol from 97 to 318 mg/day. VLDL-cholesterol decreased by 17%, LDL-cholesterol increased by 12%, and HDL-cholesterol remained constant. There was no significant change in total serum cholesterol. However, individual changes in LDL ranged widely, from −22% to +42%. This study shows the importance of individual variability to dietary therapy and thus makes general conclusions and recommendations difficult.

Increasing dietary cholesterol from 180 ± 110 to 1470 ± 80 mg/day by feeding six eggs to seven healthy normocholesterolemic young adults (three males and four females) increased LDL-cholesterol by 40% and HDL cholesterol by 12% (Nutr. Rev., 1985). Apob-100 synthesis increased by 23% whereas the fractional catabolic rate decreased by 10%. There were no changes in non-receptor-mediated LDL clearance.

In a study of eight men and one woman, seven of whom were hyper-lipidemic, increasing dietary cholesterol from 250 to 750 mg/day decreased synthesis of cholesterol but increased hepatic secretion (Nutr. Rev., 1985). In five of nine subjects in whom cholesterol intake inhibited cholesterol bio-synthesis, there was no change in serum cholesterol level. These workers found that an increase in dietary P/S ratio tended to block the increase in serum LDL-cholesterol.

It is obvious that dietary manipulation to regulate serum cholesterol levels show great variability among individuals because of the unique genetic makeup of each person. The general conclusion, however, is that consumption of three to six eggs per day can increase serum LDL- and HDL-cholesterol levels in about 50% of normocholesterolemic subjects and that this increase can be blunted by increasing the dietary P/S ratio to 1.0 or higher. This can be done easily by ingesting polyunsaturated fatty acids found in vegetable oils and fish and de-creasing the intake of foods rich in saturated fatty acids, such as butter, coconut oil, chocolate, pie crust, cream, and cakes.

How do polyunsaturated fatty acids lower blood cholesterol levels? Al-though the answer is not entirely clear, research has indicated that polyunsatu-rated fatty acids (1) decrease the synthesis of saturated fatty acids from acetate in the liver, (2) increase the conversion of cholesterol to bile acids, (3) may modify the physical properties of the lipoproteins so that they are more readily degraded by LPL, and (4) decrease the fluidity of cell membranes allowing the LDL receptor to function more effectively and endocytosis to occur more easily.

Of all foods that humans consume, egg yolks contain the highest content of cholesterol. Other foods high in cholesterol are dairy products, organ meats such as liver, kidney, and brain, and some shellfish. Cholesterol is found in all animal cells but is absent in plants and vegetables, where it is replaced by other sterols such as sitosterol, stigmasterol, and ergosterol. It is noteworthy that sitosterol and ergosterol are very poorly absorbed by the intestines, whereas cholesterol is absorbed quite efficiently. Indeed, plant sterols inhibit the absorption of cholesterol. Therefore, vegetables not only provide polyunsaturated fatty acids, which help the body convert cholesterol to bile acids, but they also provide sterols, which inhibit cholesterol absorption. High-vegetable diets therefore are cholesterol-lowering diets.

Alcohol affects nearly every organ of the body, the amount and duration of consumption being the two important variables. Excessive chronic intake of alcohol leads to fatty liver and can cause cirrhosis of the liver (see Chapter 3).

How does alcohol consumption influence blood lipids and heart disease?

The relationship is complex since relatively high rates of CHD occur in both drinkers and nondrinkers. However, lowest rates are found in moderate drinkers. The other, more recent significant finding with respect to moderate alcoholic intake is the effect on raising HDL levels. How moderate consumption raises the HDL level is not known, but the increase in HDL level may explain in part this protective effect against CHD, since HDL-cholesterol is inversely associated with CHD mortality. The possible beneficial effect of modest alcohol intake on increasing HDL levels must, however, be weighed against possible harmful effects on increasing hypertension and increasing plasma VLDL level.

It would not be prudent to recommend that teetotalers or people susceptible to alcoholism drink alcohol in order to increase their HDL levels. Moreover, the social and economic aspects in a society in which one or two glasses of wine (or one or two beers) per day are recommended by the medical profession to reduce the risk of heart disease are serious considerations that warrant very careful study. Alcohol metabolism is discussed in Chapter 3.

7.1.8. Exercise and Heart Disease

Population studies show that moderate exercise (such as uninterrupted fast walking for 30 min/day) appears to offer protection from heart disease and decreases the incidence of mortality. Moderate sustained exercise by walking daily avoids the stress on joints and the heart that can be produced by long jogging. The beneficial effect of exercise on mortality was demonstrated in studies comparing rural mail clerks with rural delivery mailmen: mortality was lower in walking mailmen. Recent studies clearly show that bus drivers have a much higher incidence of heart attacks than do men in the general population. This increased susceptibility to heart attacks is attributable in part to stress on the job and in part to lack of exercise. Blood stasis and susceptibility to clot formation by long periods of sitting down while driving may also be involved. Stress is important since it increases blood pressure (hypertension), which is a risk factor for atherosclerosis.

Prolonged moderate exercise also increases collateral circulation in the heart. This effect was demonstrated in dogs in which the circumflex coronary artery was narrowed experimentally by ligature. The dogs that were exercised on a treadmill for 15–20 min four times daily, 5 days a week for 6–8 weeks had improved collateral circulation compared with the control animals. Increased collateral circulation provides more blood to the heart.

7.1.9. Fad Diets and Other Dietary Agents

Lecithin (nicknamed "big L"), choline, and inositol are being promoted by certain books on nutrition as agents that can help alleviate or cure a number of diseases. The claims made for these agents are not supported by compelling scientific evidence. What gave rise to the use of these dietary components, and how are they supposed to function? Lecithin is a choline-phospholipid (called phosphatidylcholine) that occurs in all living cells. It is a major component of egg yolk and animal meats. Inositol and choline are biochemical compounds that can be derived from certain phospholipids. Choline is derived from the hydrolysis of lecithin, and inositol is derived from the hydrolysis of phosphatidylinositol. These phospholipids have several functions in the body. They are required for the synthesis of cell membranes, and they are components of lipoproteins that carry cholesterol and triacylglycerols in the blood. Because of their ability to interact with cholesterol and triacylglycerols they help solubilize these lipids in lipoproteins. For this reason, inositol, choline, and lecithin have been called lipotropic agents. Rats given agents such as alcohol develop fatty livers because of an excess accumulation of triacylglycerols in the liver. Fatty livers can be harmful, since they predispose the liver to hepatitis and eventually to cirrhosis. Fatty livers in rats can be ameliorated by intake of diets rich in lecithin, choline, or inositol. However, these agents have not been shown to be helpful in ameliorating fatty liver in humans. Humans, unlike rats, can produce sufficient choline and inositol to meet the body's needs. Therefore, until scientific evidence is available which clearly shows that these lipotropic agents are of benefit, they must be considered to be of little or no help for the treatment of diseases in humans. These agents do not reduce blood cholesterol levels in humans and do not protect humans from heart disease.

7.1.10. Excessive Vitamin Intake

In regard to lipid metabolism in humans, the fat-soluble vitamins A, D, and E have been recommended as dietary supplements. Vitamin E, an antioxidant, protects against peroxidation of polyunsaturated fatty acids. But how much vitamin E is required, and do we get enough in a well-balanced diet? Most vitamins function as coenzymes, i.e., they act by helping enzymes. Since enzymes and coenzymes are catalysts, they are required in very small amounts in the body. Vitamin D, however, acts more like a hormone than a typical vitamin. It is noteworthy that vitamins are not foods. Rather, most vitamins help enzymes metabolize the foods we eat. On this basis, it is not necessary for humans to take large doses of vitamins. Some vitamins, such as vitamin B12, are needed only in microgram amounts per day, whereas others, such as vitamin C, are needed in milligram amounts. The reasons for this large difference in amounts required for normal growth are related to absorption rate, the rapidity with which vitamins are lost from the body, how vitamins act as catalysts, and how effectively they are stored. Fat- soluble vitamins are stored more effectively than water-soluble vitamins, especially in the liver and fat cells.

Since 1 mg is 1000 times as much as 1 μg, the human body requires about 10,000 times as much vitamin C as it does vitamin B12. Apparently, more vitamin C is required because it is much more rapidly lost from the body.

Although most vitamins act as coenzymes and their functions have been well characterized at the molecular level, the exact mechanisms of action of vitamins E and C are not known. Vitamin A is incorporated into the pigment of the eye (rhodopsin), where it functions in vision. Vitamin A (made from dietary carotene) is also an antioxidant. Vitamin C also has antioxidant properties and helps the white blood cells kill bacteria. Only vitamin E has been shown to be effective in protecting the body from lipid peroxidation. Claims that vitamins E and C reduce cholesterol levels in humans remain to be proven.

Excessive intake of certain vitamins, in particular A and D, has been shown to be harmful. Excess amounts of vitamin D can lead to bone loss and to hypersensitivity to sunlight. Excessive intake of vitamin A can cause liver damage, which results from an alteration of the structure of cell membranes. Excessive intake of vitamin C can lead to kidney stones since vitamin C is metabolized to oxalic acid which combines with calcium to form an insoluble salt leading to calcium oxalate stones. These harmful effects of vitamins taken in excess are called hypervitaminoses.

The one vitamin that is currently being prescribed in relatively large amounts (1–6 gms/day) to lower blood lipids in humans with hyperlipidemia is niacin. This topic is discussed in Chapter 8.

7.2. BIOCHEMICAL EFFECTS OF POLYUNSATURATED FATTY ACIDS

The previous section discussed general aspects of dietary treatment of the hyperlipidemias. This section covers in more detail studies on the biochemical effects of dietary fatty acids in the treatment of hyperlipidemias and on platelet function, particularly platelet aggregation. The effects of dietary fatty acids in the treatment of hyperlipidemia and thrombosis are covered in an excellent review article by Goodnight et al. (1982). The material that follows is modified from their article.

The effects of dietary fat on atherosclerosis may be mediated by several processes. Dietary fat can alter the lipid composition of lipoproteins, which in turn can modify the enzymatic catabolism of these lipoproteins. The altered lipoproteins, especially the LDL, may interact differently with the arterial wall. Dietary fat affects the infiltration of lipids into the arterial wall by elevating or lowering the concentrations of the circulating HDL, LDL, and VLDL, which are the major transport vehicles for cholesterol and triacylglycerol in the blood. Atheromatous lipid deposits have been shown to occur when the plasma levels of

LDL-cholesterol are high. The altered lipoproteins also may influence platelet aggregation and enhance the formation of platelet thrombi.

Since 1952, the plasma cholesterol-lowering effects of polyunsaturated fat in the human diet have been demonstrated by many investigators. Polyunsaturated fat has been substituted for saturated fat in many American diets. After several weeks of feeding, many persons exhibit a lowering of plasma cholesterol levels that can be promptly reversed if saturated fat is reintroduced into the diet. *Gram for gram, saturated fat is twice as effective in raising the plasma cholesterol as polyunsaturated fat is in lowering it.* Although increased fecal excretion of cholesterol and its derivatives usually accompanies the fall in plasma cholesterol levels, the precise mechanisms of action of polyunsaturated fats remain the object of intensive study.

Suggestions that dietary fat might affect thrombosis date back to studies conducted over 20 years ago in which saturated fat was considered to be "thrombogenic" and polyunsaturated fat "nonthrombogenic." Only recently, with the discovery of platelet and vessel wall prostaglandins, has a more precise mechanism of how dietary fatty acids might affect thrombosis been elucidated. Twenty-carbon fatty acids of both the ω-6 and ω-3 families serve as substrates for the synthesis of different prostaglandins having diverse biological activities. Of particular interest is that the thromboxanes, especially TXA_2, induce platelet aggregation and vasoconstriction, whereas the prostacyclins inhibit platelet aggregation and produce vasodilation. Platelet aggregation plays an important role in atherogenesis.

7.2.1. Types of Polyunsaturated Fatty Acids

The three major classes of unsaturated fatty acids are the oleic acid (ω-9) family, the linoleic acid (ω-6) family, and the linolenic acid (ω-3) family. The structures of the fatty acids are given in Chapter 1. It is noteworthy that these fatty acid families are not metabolically interconvertible.

Polyunsaturated fatty acids are ultimately derived from plants, seeds, leaves, and phytoplankton. Seeds and leaves are the principal sources of linoleic and linolenic acids, whereas phytoplankton synthesize eicosapentenoic acid and docosahexenoic acid, the major ω-3 acids. Since phytoplankton are at the bottom of the marine food chain, all other forms of marine life eventually become enriched with these ω-3 fatty acids. Indeed, the ω-3 fatty acids may provide the degree of unsaturation required to allow fish membranes to remain fluid in very cold water.

Members of the ω-6 and ω-3 families are regarded as EFA, since they cannot be synthesized de novo in the body (see Chapter 2). The principal and best-characterized EFA is the 18:2 ω-6 linoleic acid. A dietary intake of 1–2% of total calories as linoleic acid is sufficient to prevent EFA deficiency, whereas higher levels (2–4% of total calories) are needed to reverse the condition. Linoleic acid is the obligatory precursor of arachidonic acid, which serves as a substrate for prostaglandin type 2 synthesis. This relationship between linoleic acid, arachidonic acid, and prostaglandins has led to much speculation that EFA deficiency is, in fact, a "prostaglandin deficiency." Giving certain prostaglandins to deficient animals does not uniformly alleviate the symptoms of EFA deficiency. However, since polyunsaturated fatty acids are converted to a large number of different types of prostaglandins, leukotrienes, and lipoxins, it is next to impossible to replace all of these agents in the diet in experimental studies.

The essentiality of the ω-3 family also has been questioned because of the finding that some, but not all, of the symptoms of EFA deficiency can be reversed by linolenic acid. Recent studies, however, have raised the possibility that ω-3 fatty acids may be essential not solely for prostaglandin synthesis but also for the synthesis of leukotrienes and lipoxins and as vital structural components of specialized membranes. For example, docosahexanoic acid appears to play a role in the proper functioning of nervous tissues, and impaired maze-running abilities have been found in rats deficient in ω-3 fatty acid. Remarkably high levels of docosahexenoic acid are found in brain and retinal cells as well as in spermatozoa. Although further studies will be required to document structural or metabolic defects resulting from dietary ω-3 fatty acid deprivation, the presence of high levels of these unique acids in certain tissues and the known inability of mammalian cells to synthesize them de novo imply an important dietary requirement for ω-3 fatty acids.

When dietary intakes of ω-6 and ω-3 fatty acids are adequate, little or no elongation of oleic acid to eicosatrienoic acid occurs. However, severe restrictions in dietary intake of ω-6 and ω-3 fatty acids or inhibition of fatty acid mobilization from adipose tissue can lead to the formation of significant amounts of 20:3 ω-9 fatty acid. An elevated ratio of 20:3 ω-9 to 20:4 ω-6 fatty acid in the plasma lipids or a depression of linoleic acid in plasma cholesterol esters and phospholipids is indicative of an EFA deficiency.

7.2.2. ω-6 Fatty Acids and Plasma Lipids

The hypocholesterolemic effects of polyunsaturated fatty acids present in certain vegetable and fish oils has been appreciated for almost 30 years. Not only the amount of fat in the diet but also the type (animal or vegetable) affects plasma cholesterol levels. Later studies showed that ingestion two vegetable fats (coconut oil and corn oil) led to elevated plasma cholesterol levels. The total unsaturation of the oil, rather than its source, was found to be strongly correlated with the degree of cholesterol lowering. About the same time, it was shown that the addition of polyunsaturated fat to a fat-free diet resulted in a reduction in plasma cholesterol levels that was of the same magnitude as the increase observed when saturated fats were added. The possibility that hypercholesterolemia resulted from EFA deficiency was discounted when it was shown that a fish oil high in ω-3 polyunsaturates but low in linoleic acid effectively reduced plasma cholesterol levels. On the basis of this and similar studies, researchers developed predictive equations from which the magnitude of change in plasma cholesterol that occurs after substitutions in the cholesterol or fatty acid composition of the diet could be calculated. Both equations predicted that, on a gram-for-gram basis, saturated fatty acids would raise cholesterol levels about twice as much as polyunsaturated fatty acids would lower them. These equations were derived from studies in which linoleic acid was the polyunsaturated fatty acid fed.

Several studies have demonstrated that polyunsaturated fatty acids derived from vegetable oils (principally linoleic acid) can reduce plasma cholesterol levels in normal, healthy volunteers. Typically, vegetable oils made up 40% of the total daily calories, and the cholesterol intake was about 400 mg/day. The average decline in plasma cholesterol levels in these studies was about 13% but ranged from 0 to 28%.

The extent of cholesterol lowering was related to at least two factors: (1) the magnitude of difference between the P/S ratio (as a percentage of total calories) of the saturated and the polyunsaturated diets and (2) the presence of cholesterol in the diet. For example, in eight of these studies, the P/S ratio of the polyunsaturated diet averaged about sevenfold higher than that of the saturated diet (1.5 versus 0.2), and these studies reported an average decrease in plasma cholesterol levels of only 8%. In six other studies, the P/S ratio of the polyunsaturated diet averaged 27-fold higher than that of the saturated diet (4.6 versus 0.17), and this difference was associated with an average cholesterol decrease of 20%.

It appears that plasma cholesterol levels are less responsive to changes in the dietary P/S ratio when cholesterol is absent from the diet. The relatively small decrease in plasma cholesterol levels (8%) induced by the more practical shift in P/S ratios (sevenfold) may indicate that simply changing the P/S ratio of the dietary fat without regard to other factors (e.g., the total amount of fat in the diet) may not produce physiologically important changes in the plasma cholesterol levels.

In hyperlipidemic patients, the effects of polyunsaturated vegetable oils have been similar to and, in some cases, greater than those reported in normal subjects. The average percent decrease was somewhat higher than that found in normal subjects (13–33%). However, even in these hypercholesterolemic patients, those fed diets with the smallest difference in P/S ratios (2.0 versus 0.2) experienced the smallest decreases in plasma cholesterol levels (13%).

All phenotypic forms (types I–V) of hyperlipidemia showed some hypolipidemic response to dietary polyunsaturated fats. The average declines in plasma cholesterol levels was 19% in 17 type IIa patients, 21% in 15 type IIb patients, 15% in 29 type IV patients, and 16% in 7 type V patients. Thus, no particular phenotype was unusually responsive to individual effects of saturated versus polyunsaturated fats on LDL metabolism. The fractional catabolic rate of LDL was significantly higher from the polyunsaturated diet than from either the monounsaturated or saturated regimen. This finding suggest that increased catabolism of LDL results primarily from the addition of polyunsaturated fat to the diet rather than from the removal of saturated fatty acids.

7.2.3. Membrane Fatty Acid Composition and Lipoprotein Metabolism

Ingestion of polyunsaturated fat alters the fatty acid composition of cellular membranes and plasma lipoproteins, with a resultant increase in fluidity. LPL may be more reactive toward the more fluid lipoprotein particles, which suggests that linoleate-enriched lipoproteins may be cleared more rapidly. In addition, the fatty acid composition of cellular membrane phospholipids can influence the activity of membrane-bound enzymes. An increase in the unsaturation of membrane phospholipids in cultured human fibroblasts has been found to lead to a more rapid removal of LDL particles but did not change the number of LDL receptors. Thus, the activity of receptors or membrane-bound enzymes toward lipoproteins may be enhanced by higher levels of linoleic acid in the diet.

In summary, no single mechanism of action of ω-6 polyunsaturated fatty acids in reducing blood lipids has yet been proven. Many of the observed effects of polyunsaturated fatty acids may be interrelated. For example, increases in fecal sterol excretion could be secondary to an increased rate of catabolism of LDL. This effect may be caused by changes in lipoprotein or membrane fluidity, leading to changes in the activity of lipolytic enzymes or membrane receptors.

7.2.4. ω-3 Fatty Acids and Plasma Lipids

Studies by Bang et al. (1980) stimulated interest in the metabolism of ω-3 fatty acids and suggested a link between the habitual ingestion of these fatty acids in the diet and the low death rates from atherosclerotic disease observed in Greenland Eskimos. Although few reliable data on cardiovascular morbidity and mortality among this particular group of Eskimos are available, Bang et al. noted that from 1963 to 1967, only two cases of atherosclerotic heart disease and no cases of diabetes mellitus were registered in the population of 1300 people. Estimated deaths from cardiovascular disease were only 16% of all deaths in Alaskan coastal Eskimos (ages 41–70 years) from 1959 to 1968. Bang et al. have provided evidence that the high levels of ω-3 fatty acids consumed in the Eskimo diets of seal, whale, and fish (5.8 g/day) may be responsible for the lower plasma lipid levels in this population compared with levels in Danes (0.8 g/day). In addition, ω-3 fatty acids may also be responsible for the prolonged bleeding times that are common in this group of Eskimos.

Between 1956 and 1963, eight studies of fish oil feeding were carried out in humans. Although these studies varied in the kinds of subjects studied, the degree of dietary control, the sources of fish oil fatty acids, and duration, there was good agreement that fish oils were at least as hypocholesterolemic as polyunsaturated vegetable oils. Two interesting features of these studies were perhaps not appreciated at the time. First, since fish oils contain cholesterol (300–500 mg/dl) and vegetable oils do not, the majority of these investigators fed higher levels of cholesterol during the fish oil phase than during the vegetable oil phase. Thus, the finding of similar hypocholesterolemic effects of these two types of fat implied that the fish oils might have been even more hypocholesterolemic if they had not contained dietary cholesterol.

Second, the daily intake of ω-6 fatty acids was invariably greater than the intake of ω-3 fatty acids, yet similar or greater reductions in plasma cholesterol levels occurred from ω-3 fatty acids. Thus, gram for gram, the ω-3 fatty acids were considerably more hypocholesterolemic than ω-6 linoleic acid. More recently, dramatic reductions in the concentrations of plasma triacylglycerols and VLDL have been observed in both normolipidemic and hyperlipidemic subjects fed diets supplemented with fish oils. In one study, normal subjects were fed diets enriched with salmon oil, polyunsaturated vegetable oils, and a relatively saturated control diet for periods of 4 weeks each. Plasma cholesterol levels were reduced similarly with the salmon and vegetable oil diets. In contrast, plasma triacylglycerol levels fell 33% as a result of the salmon oil diet but were unchanged after the vegetable oil diet. This direct comparison provided the first clear demonstration that ω-3 fatty acids present in salmon oil possess unique hypotriglyceridemic properties not found in the polyunsaturated ω-6 fatty acids of vegetable oils. These studies further demonstrated that dietary ω-3 fatty acids significantly decreased plasma VLDL (50%) and LDL (16%) levels but did not change the concentrations of HDL. The fact that HDL levels were not depressed has been confirmed by others; in some studies, HDL-cholesterol levels have actually increased on the ω-3 fatty acid-rich diets.

The effects of fish oils in hypercholesterolemic and hypertriglyceridemic patients have been reported. Plasma cholesterol and triacylglycerol levels were both reduced more by menhaden fish oil than by corn oil. In patients with type IIb hyperlipidemia, plasma cholesterol levels decreased from 294 to 239 mg/dl on the ω-6 vegetable oil diet and to 200 mg/dl on the ω-3 diet. Plasma triacylglycerol levels were 397 mg/dl on control diets, 247 mg/dl on the vegetable oil diet, and 135 mg/dl on the salmon oil diet. Thus, the fish oil diet was significantly more hypolipidemic than the polyunsaturated vegetable oil diet.

The effects of dietary fish oil on the plasma lipids of patients with severe type V hypertriglyceridemia were first reported in 1981. These patients, in whom hypertriglyceridemia is normally exacerbated by even moderate intakes of dietary fat, were able to consume up to 30% of their calories as fish oil, with a concomitant lowering of plasma lipid levels. The ω-3 fatty acid-rich diet resulted in a 46% decline in plasma cholesterol levels and a 77% drop in triacylglycerol levels. Comparative diets containing 30% of calories as ω-6 fatty acids caused rapid increases in the concentrations of both plasma cholesterol and triacylglycerols; consequently it became necessary to stop these diets after 14 days because of the increasing risk of abdominal pains, hepatomegaly, and acute pancreatitis. The ω-3 fatty acids had effects on plasma lipid levels distinctly different from those of ω-6 fatty acids, and thus they may be of substantial therapeutic benefit in hypertriglyceridemic patients.

Attempts have been made to determine the minimum intake of ω-3 fatty acids necessary to produce significant changes in plasma lipid levels. Although dose responses have not been fully ascertained, it was shown that approximately 4 g/day of ω-3 fatty acids from 20 ml of cod liver oil produced a greater cholesterol lowering than did 16 g of linoleic acid. In a second study, however, the addition of up to 8 g of ω-3 fatty acids (from cod liver oil) did not change the plasma cholesterol levels but did result in a significant reduction in plasma triacylglycerols. Another study showed that 4 g/day of ω-3 fatty acids could reduce the levels of plasma triacylglycerols but not of cholesterol. A major problem with many of these studies is that fish oils were given simply as supplements and were not incorporated into a metabolic diet as a replacement for a portion of the customary dietary fat. In addition, the cholesterol intake was usually higher during the fish oil phase. It seems likely that more significant hypocholesterolemic effects might have occurred if the dietary periods had been more rigorously matched in terms of both fat and cholesterol content.

In summary, fish oils have always been either as effective as or more effective than vegetable oils in lowering plasma cholesterol levels. These reductions occurred when ω-3 fatty acids constituted 1–8% of total calories. In light of the fact that linoleic acid intakes of 15–20% of calories were needed to achieve similar depressions in plasma cholesterol levels, the dietary ω-3 fatty acids were roughly two to five times more potent than the ω-6 acids.

The most striking effects of fish oil feeding have been the rapid and drastic decreases in the levels of plasma triacylglycerol and VLDL. Since vegetable oils produce little depression in plasma triacylglycerol concentrations, it is evident that ω-3 fatty acids have unique effects not shared with the ω-6 fatty acids. Studies in hyperlipidemic subjects have demonstrated that small quantities of ω-3 fatty acids are much more hypolipidemic than large amounts of linoleic acid. Thus, dietary ω-3 fatty acids may prove to be very effective in the treatment of hyperlipidemia.

7.2.5. Hypolipidemic Mechanisms of the ω-3 Fatty Acids

Data on ω-3-induced hypocholesterolemia suggest that the following mechanisms may be involved: (1) increased fecal sterol excretion, (2) changes in the fatty acid composition (fluidity) of lipoproteins, and (3) changes in the rates of synthesis and/or catabolism of VLDL and LDL. It also is likely that ω-3 fatty acids inhibit the activity of acetyl-CoA carboxylase and the fatty acid synthase complex thus inhibiting the de novo synthesis of fatty acids in the liver.

Although relatively little data on the fatty acid composition of individual lipid classes from the major lipoprotein fractions have been reported, the fatty acids of each plasma lipid class have been studied. The percentage of total fatty acids occurring as ω-3 in each class of plasma lipids (triacylglycerol, phospholipid, and cholesterol esters) was 1–3% on the control diet and increased to 31, 33, and 26%, respectively, after the ω-3 diet. Since these ω-3 fatty acids are highly unsaturated, the fluidity of the lipoprotein particles would be affected and the interaction between lipoproteins and lipolytic enzymes may be enhanced.

7.3. PLATELET AND BLOOD VESSEL PROSTAGLANDINS DERIVED FROM DIETARY POLYUNSATURATED FATTY ACIDS

An understanding of the relationships of polyunsaturated fatty acids to their role in platelet and vascular function has been greatly aided by the recent identification and characterization of platelet and endothelial cell prostaglandins. Since these potent substances are ultimately derived from polyunsaturated fatty acids in the diet, it is possible that manipulation of dietary fat content can alter prostaglandin synthesis and subsequent platelet–vessel interactions.

Direct uptake of free arachidonic acid by platelets or the vessel wall leads to its incorporation into membrane phospholipids. Following activation of one or more phospholipases, arachidonic acid becomes available to the enzyme cyclooxygenase and is rapidly converted to the labile cyclic endoperoxides. Enzymes such as thromboxane synthetase in the platelet or prostacyclin synthetase in the endothelial cell convert the endoperoxides to biologically active TXA_2 or PGI_2. The potent vasoconstricting and platelet-aggregating effects of TXA_2 and the vasodilating and platelet inhibitory effects of PGI_2 are now well known.

The EFA linoleic acid is converted to to arachidonic acid in the liver. Although the ingestion of diets rich in linoleic acid regularly leads to accumulation of this acid in platelet membrane phospholipids, arachidonate levels are found to be unchanged or even decreased in platelets. Similarly, feeding of linolenic acid does not appear to lead to significant increases of eicosapentenoic acid in adult human plasma. The feeding of marine food rich in eicosapentenoic acid, however, leads to rapid incorporation of this fatty acid into both platelet and endothelial cell membranes. Feeding ω-3 fatty acids to humans leads to enrichment of these acids in the alkenylacyl-phosphatidylethanolamines of platelet membranes (Aukema and Holub, 1989).

Arachidonic acid can also be converted by the lipoxygenase pathway to a class of compounds called leukotrienes and lipoxins. Since some leukotrienes have been shown to have bronchoconstrictive properties and chemotactic activity and possibly to influence thromboxane and prostacyclin synthesis, they will provide a fertile area for future biochemical and clinical research.

7.4. EFFECTS OF POLYUNSATURATED FATTY ACIDS ON PLATELET FUNCTION

7.4.1. Linoleic Acid

Early studies of fatty acids and thrombosis generally supported the hypothesis that replacement of saturated by polyunsaturated fatty acids in the diet might be protective against thromboembolic events and that this protective effect might be mediated through an inhibition of platelet function. Consequently, during the last decade a number of investigators have examined the effects of diets containing large amounts of polyunsaturated fatty acids (especially linoleic acid) on platelet lipid composition and hemostatic function. Evaluation of these studies has been somewhat difficult because of marked variations in experimental design, dietary control, lipid analyses, and tests of platelet function. Nonetheless, some important generalizations can be made. In some experimental situations, ingestion of a diet very rich in linoleic acid may lead to a selective decrease in arachidonic acid, whereas intake of a diet containing saturated fatty acids may produce the opposite effect. Whether these changes in arachidonic acid content translate into equivalent changes in thromboxane or prostacyclin release remains to be determined.

Changes in platelet function have been studied after the chronic feeding of diets rich in polyunsaturated vegetable oils. When bleeding times of human subjects ingesting a diet rich in ω-6 polyunsaturated fatty acids were compared with results for a control dietary period, no significant differences were found. Despite this lack of prolongation in the bleeding times, some tests of platelet aggregation have been affected by dietary linoleic acid.

It appears that supplementing or replacing saturated fat with oils rich in linoleic acid leads to mild and rather variable effects on platelet function. Although the bleeding time is not significantly prolonged, platelet aggregation by low doses of aggregating agents may be inhibited, platelet retention on glass beads may be reduced, and the time for platelet aggregation in the filtragometer may be prolonged. Platelets also appear to be less reactive as evidenced by lengthening of the thrombin clotting time and reduction in circulating platelet aggregates. Studies indicate that only the feeding of fish oil gives a prolonged bleeding time.

7.4.2. Linolenic Acid

In certain tissues, linolenic acid may be desaturated and elongated to eicosa-pentenoic acid. Feeding of oils rich in linolenic acid (e.g., linseed oil, which contains 53% linolenic acid) to animals or humans might lead to the accumulation of eicosapentenoic acid in tissues such as platelets or endothelial cells. One such study showed that feeding rats up to 4% of calories as purified methyl linolenate led to a decrease in arachidonic acid (from 23 to 11%) and an increase in eicosapentenoic acid from 0.1 to 3–4% in liver and serum lipids. In the few studies that have examined the effects of feeding linolenic acid upon platelet or endothelial cell function, no changes were found in platelet adhesiveness or bleeding times in humans after dietary supplements of linseed oil.

7.4.3. Eicosapentenoic Acid

As mentioned earlier, the fascinating observations by Bang et al. (1980) generated provocative questions about the relationships of an "Eskimo diet" to platelet and vascular function. Previous work had established that the Eskimo diet contained large amounts of eicosapentenoic acid and that ω-3 fatty acids were present in plasma and platelets of the Eskimos. A bruising or bleeding tendency among the Greenland Eskimos had been described and was also noted in the 1930s by a French explorer who visited northern Canadian Eskimos accustomed to eating a similar diet. Pursuing these leads, the Danish investigators studied the bleeding times of their Eskimo subjects and found them to be prolonged in comparison with results for a control population. These and other observations have stimulated a growing interest in the effects of ω-3 fatty acids derived from marine oils on the composition and function of platelets and on cellular prostaglandin metabolism.

7.5. PLATELET LIPID COMPOSITION

The concentrations of both linoleic acid and arachidonic acid were decreased in the Eskimo platelet phospholipids. Similar observations have been noted in volunteers fed mackerel oil, cod liver oil, or salmon oil and in animals given menhaden or cod liver oil. This reduction in linoleic acid could be a reflection of the low linoleate intake (most fish oils contain only 1–2% linoleic acid) or may relate to the displacement of linoleate by other fatty acids. It is possible that the decreased dietary linoleic acid may lead to reduced arachidonic acid levels in plasma and in platelet membrane phospholipids. As might be expected, the consumption of ω-3-rich fish oils led to a marked incorporation of eicosapentenoic acid in the platelet membrane phospholipids.

7.6. PLATELET AND VASCULAR FUNCTION

Both platelet function and platelet–vessel wall interactions have been altered by diets enriched in ω-3 fatty acids. With one exception, bleeding times have invariably been prolonged by 30–40% in humans ingesting fish oils for several weeks or longer although a clinical bleeding tendency has not been observed.

Decreased platelet counts have been observed in subjects ingesting large amounts of ω-3 fatty acids. These findings include the early Eskimo observations as well as results of a recent study of the effects of salmon oil feeding in volunteers. In general, the reduction in platelet count, while statistically significant, remained within the normal range and was not low enough to affect the duration of bleeding times. In several individuals, however, there was a more marked fall in the platelet count. After the salmon oil feeding was discontinued, the platelet count rose rapidly to normal levels. In one of these subjects, a repeat study in which the subject was fed salmon oil again led to a reduction in platelet count although platelet survival was unchanged. At present, it is not clear whether this thrombocytopenia is related to the ω-3 fatty acids in the fish oil, to other natural components, or to contaminants.

In summary, feeding of ω-3 fatty acid-rich fish oils to humans leads to a reproducible prolongation of the Ivy bleeding time, inhibition of platelet aggregation by ADP and collagen, and a decrease in platelet retention on glass beads. In some cases, there may also be a reduction in platelet count. The mechanisms for these functional alterations may be explained in part by changes in platelet and endothelial cell prostaglandin synthesis induced by alterations in dietary fatty acid composition.

7.7. MECHANISMS OF FATTY ACID EFFECTS

Some studies in which fish oils were fed to humans showed production of thromboxane B_2 (the stable metabolite of TXA_2) was decreased after stimulation of platelets by collagen or ADP.

The production of PGI_2-like activity from vessel walls after fish oil feeding has also been examined. Incubated segments of rat aorta produced less platelet inhibitory activity (presumed to be PGI_2) in animals ingesting cod liver oil compared with sunflower seed oil. It seems clear that feeding of fish oils containing ω-3 fatty acids leads to a reduction in TXA_2 production by platelets, and there may also be a reduction in PGI_2 but an increase in PGI_3 from the walls of blood vessels.

One explanation for the reduced formation of thromboxane B_2 from platelets enriched in eicosapentenoic acid may involve competitive inhibition of cyclooxygenase by eicosapentenoic acid which leads to reduced TXA_2 synthesis from arachidonic acid. In addition, eicosapentenoic acid may be a relatively poor substrate for cyclooxygenase; consequently only very small amounts of TXA_3 are produced. It is also possible that prostaglandins of the D series (e.g., PGD_3) may be formed from eicosapentenoic acid and that these may also inhibit platelet aggregation. More importantly, however, is the fact that eicosapentenoic acid produces TXB_3 in platelets and PGI_3 in endothelial cells (see pp. 142–143).

Other hypotheses for the decreased platelet responsiveness that follows ω-3 fatty acid ingestion are possible: blockade of thromboxane receptors on platelets by eicospentenoic acid; replacement of arachidonic acid by eicosapentenoic acid in platelet phospholipids, with a concomitant reduction in the amount of arachidonate available for TXA_2 synthesis; or an inhibitory effect of eicosapentenoic acid on phospholipase A_2, leading to a reduced release of arachidonic acid from platelet phospholipids.

The effects of eicosapentenoic acid on prostaglandin production by endothelial cells have received less attention. Preliminary evidence indicates that both PGI_2 and PGI_3 are decreased, although some platelet inhibitory activity may still be present. Since bleeding times are prolonged in humans fed fish oil, it may be that there is a relatively greater decrease in thromboxane generation in platelets than in prostacyclin production in endothelial cells.

In conclusion, ingestion of dietary fish oils containing eicosapentenoic acid may have profound effects on platelet or vessel composition and function. Cellular phospholipid concentrations of arachidonic acid are decreased, bleeding times are prolonged, and various in vitro tests of platelet function are inhibited. One explanation for the platelet inhibition is the significant reduction in platelet TXA_2 which is replaced by TXA_3 which does not stimulate platelet aggregation (see Chapter 8).

7.8. POSSIBLE RISKS AND SIDE EFFECTS FROM INCREASED AMOUNTS OF DIETARY POLYUNSATURATED FATTY ACIDS

7.8.1. ω-6 Fatty Acids

There is no precedent attesting to the safety of a diet high in linoleic acid. For most of humankind, prior to the advent of 20th-century technology, the total fat content of the diet has ranged between 8 and 12% of the total calories and linoleic acid constituted from 4–8% of the total calories, which has been the typical intake of most Americans. These historical observations indicate that there is no evidence concerning the long-term beneficial or harmful effects of consuming diets that might contain up to 15–25% of total calories from linoleic acid. In addition, studies in present day populations consuming both high- and low-fat diets have revealed that the linoleic acid contents of adipose tissue are very similar worldwide. Since linoleic acid in adipose tissue can be derived only from the diet, this finding implies that the typical intake in America is usual around the world.

Possible harmful effects of dietary ω-6 fatty acids include enhanced formation of cholesterol gallstones, a stimulus to carcinogenesis, increased vitamin E requirement, promotion of obesity, increased uptake of plant sterols, and increased cholesterol absorption. In humans, the increased excretion of bile acids and neutral sterols in the stool after polyunsaturated fat feeding may imply a change in bile acid composition that could influence gallstone formation.

An increased incidence of malignant neoplasia in individuals who consumed a diet high in polyunsaturated fat has been reported in some studies but not in others. It is believed that an increase in neoplasia may be promoted by an increased accumulation of peroxides from linoleic acid, which may enhance the production of carcinogens or procarcinogens.

7.8.2. ω-3 Fatty Acids

Several aspects of dietary fish oils are of possible concern. Some fish oils contain high levels (about 10%) of cetoleic acid (22:1 ω-11) which is an isomer of erucic acid (22:1 ω-9). Erucic acid is found in rapeseed oil and other brassica-derived oils and, when fed in high levels is known to cause transient myocardial lipidosis and fibrosis in several experimental animals. However, erucic acid and cetoleic acid have not been shown to have detrimental effects in humans. Eskimo populations have consumed diets high in cetoleic acid for centuries, and no myocardial damage has been reported. In humans fed fish oils experimentally, cetoleic acid has not been detected in the plasma.

Another possible side effect of high levels of fish oils has been reported in pigs fed mackerel oil or whale oil. These animals developed a disorder known as yellow fat disease which is associated with vitamin E deficiency. This condition has not been reported in Eskimos or in other species of experimental animals. Because of the highly unsaturated nature of the ω-3 fatty acids, the vitamin E requirement may be increased during the consumption of fish oils as it is during intake of polyunsaturated vegetable oils. This potential problem can be prevented by taking adequate amounts of vitamin E.

Two effects of fish oil feeding on platelet function may be of concern; increased bleeding time and thrombocytopenia. A mild thrombocytopenia has been reported in most fish oil feeding studies in humans. This effect has not been a cause for concern except in two subjects who transiently developed platelet counts below 100,000 after 3 weeks of a diet containing 40% of calories as salmon oil. A few days after the salmon oil diet was discontinued, the platelet counts returned to normal. Such thrombocytopenia has not been observed in subjects receiving lesser amounts of salmon oil.

The expected benefits of modest increases in dietary polyunsaturated fat would far outweigh the possible risks outlined above. While linoleic acid intake of up to 10% of total calories would not be considered dangerous or deleterious, only slight effects on plasma lipid levels and platelet function may be expected. On the other hand, intake of ω-3 fatty acids is currently very low. Therefore, the inclusion of as much as 4–8 g per day (1–2% of calories) may have substantial effects on plasma levels, platelet function, and bleeding times without incurring significant risk.

The uncontrolled consumption of large amounts (15–50 ml/day) of fish liver oils such as cod liver oil should certainly be discouraged because of the high levels of vitamins A and D. A relatively small amount of cod liver oil (20 ml) would provide four times the government recommended dietary allowance (RDA) of each vitamin. Toxic symptoms have been reported with as little as five times the RDA of vitamin A. On the other hand, the daily consumption of 0.5 lb of salmon would provide approximately 30 g of fish oil (4–6 g of ω-3 fatty acids) and very small amounts of the fat-soluble vitamins.

At present it is reasonable to recommend eating marine fish such as salmon two or three times per week to obtain sufficient ω-3 fatty acids that can offer some protection against platelet aggregation while lowering blood cholesterol and triacylglycerol levels. This dietary regimen is particularly important for persons who have elevated blood lipids and are at greater risk for developing myocardial infarction and for individuals who have suffered a myocardial infarction. Obviously, this dietary therapy must be part of a more complete dietary regimen to lower blood lipids and to minimize thrombogenesis in high-risk individuals such as those who have had a myocardial infarction.

In a recent review, Harris (1989) notes that new trials using smaller, more practical doses of fish oil supplements have confirmed the hypotriglyceridemic effect, they have little effect on total cholesterol levels and in hyper-triglyceridemic patients they increase LDL-cholesterol levels. Discrepancies among fish oil studies regarding the effects of ω-3 fatty acids on LDL-cholesterol are attributable to experimental design. In the majority of studies reporting reductions in LDL-cholesterol levels, saturated fat intake was lowered when switching from the control diet to the fish oil diet. When fish oil is fed and saturated fat intake is constant, LDL-cholesterol levels either do not change or may increase. Levels of HDL-cholesterol are increased by about 5–10% with fish oil intake. The decrease in plasma triacylglycerol levels appears to result from an inhibition of hepatic triacylglycerol synthesis. Harris concludes that further studies will be needed to document the hypolipidemic and antiatherogenic effects of ω-3 fatty acids. It is noteworthy that for some persons the hypolipidemic effect will be important whereas for others, especially those at risk for myocardial infarction, the antithrombogenic effect will be important.

REFERENCES

Anderson, J. W., Zettwoch, N., Feldman, T., Tietyen-Clark, J., Oeltgen, P., and Bishop, C. W., 1988, Cholesterol-lowering effects of psyllium hydrophilic mucilloid for hypercholesterolemic men, *Arch. Intern. Med.*, 148:292.

Aukema, H. M., and Holub, B. J., 1989, Effect of dietary supplementation with a fish oil concentrate on the alkenylacyl class of ethanolamine phospholipid in human platelets, *J. Lipid Res.*, 30:59.

Balasubramaniam, S., Simons, L. A., Chang, S., and Hinkle, J. B., 1985, Reduction in plasma cholesterol and increase in biliary cholesterol by a diet rich in n-3 fatty acids in the rat, *J. Lipid Res.*, 26:684.

Balasubramanian, K. A., Manohar, M., and Mathan, V. I., 1988, An unidentified inhibitor of lipid peroxidation in intestinal mucosa, *Biochim. Biophys. Acta*, 962:51.

Bang, H. O., Dyerberg, J., and Sinclair, H. M., 1980. The composition of the Eskimo food in northwestern Greenland, *Am. J. Clin. Nutr.*, 33:2657.

Barlow, S. M., and Stansby, M. E. (eds.), 1982, *Nutritional Evaluation of Long Chain Fatty Acids in Fish Oil*, Academic Press, New York.

Black, K. L., Culp, B., Madison, D., Randall, O. S., and Lands, W. E. M., 1979, The protective effects on dietary fish oil on focal cerebral infarction, *Prostagl. Med.*, 5:247.

Cholesterol metabolism and ω-3 polyenes in fish oils, 1986, *Nutr. Rev.*, 44:147.

Cohen, L. A., 1987, Diet and cancer, *Sci. Am.*, 257:42.

Diplock, A. T., Balasubramanian, K. A., Manohar, M., and Mathan, V. I., 1988, Purification and chemical characterization of the inhibitor of lipid peroxidation from intestinal mucosa, *Biochim. Biophys. Acta*, 962:42.

Dyerberg, J., 1986, Linolenate-derived polyunsaturated fatty acids and prevention of atherosclerosis, *Nutr. Rev.*, 44:125.

Dyerberg, J., and Bang, H. O., 1979, Haemostatic function and platelet polyunsaturated fatty acids in Eskimos, *Lancet*, 2:433.

Ehsani, A. A., 1987, Cardiovascular adaptations to exercise training in the elderly, *Fed. Proc.*, 46:1840.

Experimental myocardial infarction and fish oil, 1981, *Nutr. Rev.*, 39:316.

Fish oil and the development of atherosclerosis, 1987, *Nutr. Rev.*, 45:90.

Gerstenblith, G., Renlund, D. G., and Lakatta, E. G., 1987, Cardiovascular response to exercise in younger and older men. *Fed. Proc.*, 46:1834.

Goodnight, S. H., Harris, W. S., Connor, W. E., and Illingworth, D. R., 1982, Polyunsaturated fatty acids, hyperlipidemia and thrombosis, *Atheroscl.*, Vol. 2, pp. 87–113.

Grundy, S. M., 1986, Comparison of monounsaturated fatty acids and carbohydrates for lowering plasma cholesterol, *N. Engl. J. Med.*, 314:745.

Hansen, W. S., 1986, The essential nature of linoleic acid in mammals, *Trends Biochem. Sci.*, 11:263.

Harris, S. H., 1989, Fish oils and plasma lipid and lipoprotein metabolism in humans: a critical review, *J. Lipid Res.* 30:785.

Hartz, A. J., Anderson, A. J., Brooks, H. L., Manley, J. C. Parent G. T., and Barboriak, J. J., 1984, The association of smoking with cardiomyopathy, *N. Engl. J. Med.*, 311:1201.

Hulley, S. B., and Dzvonik, M. L., 1984, Alcohol intake, blood lipids and mortality from coronary heart disease, *Clin. Nutr.*, 3:139.

Influence of eggs on plasma lipoproteins, 1985, *Nutr. Rev.*, 43: 263.

Kent, K. M., Bonow, R. O., Rosing, D. R., Ewels, C. J., Lipson, L. C., McIntosh, C. L., Bacharach, S., Green, M., and Epstein, S. E., 1982, Improved myocardial function during exercise and after successful percutaneous transluminal coronary angioplasty, *N. Engl. J. Med.*, 306:441.

Kritchevsky, D., 1982, Trans fatty acid effects in experimental atherosclerosis, *Fed. Proc.*, 41:2813.

Kromhout, D., Bosschieter, E. B., and De Lezenne Coulander, C., 1985, The inverse relation between fish consumption and 20-year mortality from coronary heart disease. *N. Engl. J. Med.*, 312: 1205.

Lecos, C., 1983, A compendium on fats, *FDA Consumer*, March.

Levy, R. I. and Ernst, N., 1973. Diet, hyperlipidemia, and atherosclerosis, in *Modern Nutrition in Health and Disease*, 5th edition, chapter 31, (R. S. Goodhart and M. E. Shils, eds., Lea-Febiger, Philadelphia.

Marcus, A. J., 1984, The eicosanoids in biology and medicine, *J. Lipid Res.*, 25:1511.

Marine oils and platelet function in man, 1984, *Nutr. Rev.*, 42: 189.

Mattson, F. H., and Grundy, S. M., 1985, Comparison of effects of dietary saturated, monounsaturated, and polyunsaturated fatty acids on plasma lipids and lipoproteins in man, *J. Lipid Res.*, 26:194.

Mead, J. F., 1984, The non-eicosanoid functions of the essential fatty acids, *J. Lipid Res.*, 25:1517.

Mertz, W., 1982, Trace minerals and atherosclerosis, *Fed. Proc.*, 41:2807.

Miller, R. W., 1986, Diet, exercise, and other keys to a healthy heart, *FDA Consumer*, February.

Mortality from coronary heart disease is inversely related to fish consumption in the Netherlands, 1984, *Nutr. Rev.*, 43:271.

Murry, R. K., Granner, D. K., Mayes, P. A., and Rodwell, V. W., 1988, *Harpers Biochemistry*, 21st ed.1, Appleton & Lange, Norwalk, Connecticut.

National Cholesterol Education Program Expert Panel, 1988, Report of the National Cholesterol Education Expert Panel on detection, evaluation, and treatment of high blood cholesterol in adults, *Arch. Intern. Med.*, 148:360.

Nettleton, J. A., 1985, *Seafood Nutrition. Facts, Issues, and Marketing of Nutrition in Fish and Shellfish*, Osprey Books, Huntington, New York.

Normand, F., 1987, Binding of bile acids and trace minersls by soluble hemicellulose of rice, *Food Technol.*, 41:87.

Olson, A., Gray, G.M., and Chiu, M. C., 1987, Chemistry and analysis of soluble dietary fiber, *Food Technol.*, 41:71.

Oscai, L. B., Patterson, J. A., Bogard, D. L., Beck, R. J., and Rothermell, B. L., 1972, Normalization of serum triglycerides and lipoprotein electrophoretic patterns by exercise, *Am. J. Cardiol.*, 30:775.

Pariza, M. W., 1987, Dietary fat, calorie restriction, ad libitum feeding, and cancer risk, *Nutr. Rev.*, 45:1.

Phillipson, B. E., Rothrock, D. W., Connor, W. E., Harris, W. S., and Illingworth, D. R., 1985, Reduction of plasma lipids, lipoproteins, and apoproteins by dietary fish oils in patients with hypertriglyceridemia, *N. Engl. J. Med.*, 312:1210.

Reduction of plasma lipids and lipoproteins by marine fish oils, 1985, *Nutr. Rev.*, 43:268.

Sirtori, C., Rucci, G., and Gorini, S. (eds.), 1975, Diet and Atherosclerosis, Plenum Press, New York.

Slavin, J. L., 1987, Dietary fiber:classification, chemical analysis, and food sources, *J. Am. Dietetic Assoc.*, 87:1164.

Story, J. A., 1982, Dietary carbohydrates and atherosclerosis, *Fed. Proc.*, 41:2797.

U.S. Department of Health and Human Services, 1985, Facts about blood cholesterol, Publication NIH 85-2696, U.S. Government Printing Office, Washington, D. C.

Vahouny, G. V., 1982, Dietary fiber, lipid metabolism, and atherosclerosis, *Fed. Proc.*, 41:2801.

Van Horn, L., Emidy, L. A., Liu, K., Liao, Y., Ballew, C., King, J., and Stamler, J., 1988, Serum lipid responses to a fat-modified oatmeal-enhanced diet, *Prev. Med.*, 17:377.

Willis, A. L., 1981, Nutritional and pharmacological factors in eicosanoid biology, *Nutr. Rev.*, 39:289.

Wood, P. D., Terry, R. B., and Haskell, W. L., 1985, Metabolism of substrates: diet, lipoprotein metabolism and exercise, *Fed. Proc.*, 44:358.

Zanni, E. E., Zannis, V. I., Blum, C. B., Herbert, P. N., and Breslow, J. L., 1987, Effect of egg cholesterol and dietary fats on plasma lipids, lipoproteins, and apoproteins of normal women consuming natural diets, *J. Lipid Res.*, 28:518.

Chapter 8

DRUG THERAPY FOR HYPERLIPIDEMIAS
Lipid-Lowering Drugs and Antithrombic and Fibrinolytic Drugs

Dietary management is tried first to lower plasma levels of cholesterol and triacylglycerols in patients with hyperlipidemia. Many patients find it difficult to be on a restricted and regimented diet for a long period of time and hence compliance is compromised. In this situation, drug therapy is necessary to maintain plasma lipid levels in the normal range.

8.1. LIPID-LOWERING DRUGS

When dietary therapy fails to lower plasma cholesterol to the desirable range, drug therapy is usually initiated. The drugs of choice for persons who have elevated LDL-cholesterol but normal plasma triacylglycerol levels are Colestid (or cholestryamine) and Mevacor (also called Lovastatin), alone or in combination. The drugs of choice for patients with concurrent hypertriglyceridemia (i.e., triacylglycerol level >250 mg/dl) are Colestid plus nicotinic acid (niacin) or Mevacor plus nicotinic acid. Lopid (gemfibrozil) is also effective in lowering plasma triacylglycerol levels and can be used in combination with Colestid and Mevacor. These drugs alone and especially in combination have been shown to reduce plasma cholesterol and triacylglycerol and offer protection from CHD. Many more studies must be done to determine which combination of drugs is most effective in both lowering plasma lipids and protecting against CHD. Lopid has replaced Atromid-S because of the undesirable side effects of the latter drug.

169

Structures of the drugs used for treatment of hyperlipidemias are shown in Fig. 8-1.

It is noteworthy that all of these drugs have side effects that differ among patients and some drugs are easier to take than others. The rationale for selection of the various drugs is shown in Table 8-1.

Colestid or cholestyramine is recommended for patients with elevated plasma cholesterol. These insoluble drugs are taken while suspended in fruit juices or water. The dose varies from 10 to 30 g/day, depending on the degree of hypercholesterolemia and the dosage that the patient can tolerate. They reduce plasma cholesterol (mainly LDL-cholesterol) by 13–25%. Patient compliance with these bile acid-binding resins has been low. Constipation, especially at higher doses, is an undesirable side effect.

Nicotinic acid (the vitamin niacin) in doses of 1–6 g/day is recommended for patients with elevated plasma cholesterol or triacylglycerols. (It is noteworthy that niacinamide, the amide form of niacin, is not effective). Nicotinic acid reduces plasma cholesterol and triacylglycerol by about 9–20%. Starting the patient with a low dose (100 mg/day) and increasing the dosage over a period of 1–2 weeks prevents the severe flushing that initially may occur. Giving aspirin during this initial period also helps prevent flushing.

Mevacor is recommended for patients with elevated cholesterol, especially LDL-cholesterol. Taken in doses of 10–40 mg per day, this drug lowers cholesterol (mainly LDL-cholesterol by 17–33 % and increases HDL-cholesterol by 8–11 %.

Lopid in doses of 0.6–3 g/day is recommended for patients with elevated cholesterol and triacylglycerols. It lowers plasma cholesterol by 2–9%.

Table 8-1. Drug Therapy for High-Risk LDL-Cholesterol

First drug	
Colestid and cholestyramine	Bile acid binding resins, prevent reabsorption of bile acids
Mevacor (Lovastatin)	Inhibits HMGR, inhibits cholesterol synthesis
Second drug	
Nicotinic acid (niacin)	Inhibits synthesis of VLDL, possibly by inhibiting synthesis of apoB-100; inhibits lipolysis
Alternate second drug	
Lopid (gemfibrozil)	Inhibits synthesis of VLDL, possibly by inhibiting synthesis of apoB-100
Lorelco (probucol)	Inhibits cholesterol synthesis and increases fecal excretion of bile acids may decrease plasma HDL levels

Generic name	Trademane	Structural formula	U.S. supplier	Date*
Cholestryamine	Questran	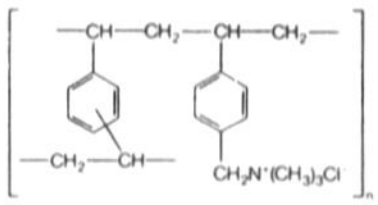	Mead Johnson	1965
Colestipol	Colestid	Complex resin similar to Questran	Upjohn	1977
Gemfibrozil	Lopid	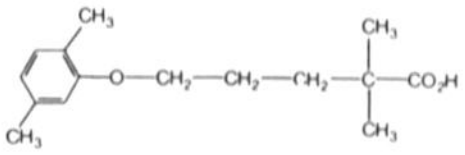	Park-Davis	1982
Nicotinic acid (Niacin)	Nicolar and others	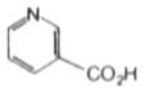	Armour and others	1955
Probucol	Lorelco	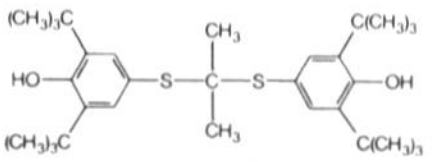	Merrell Dow	1977
Lovastatin (Mevinolin)	Mevacor	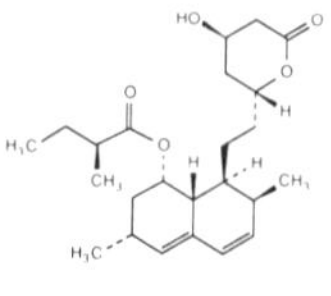	Merck Sharp & Dohme	1987

* Year of commercial introduction in the United States.

Figure 8-1. Structure of drugs used to treat hyperlipidemia. (From *Chem. Eng. News,* July 12, 1988, p. 37.)

Probucol (Lorelco) lowers plasma cholesterol. A dose of 1 g/day lowers cholesterol by 1–2.7%. This drug does not lower plasma triacylglycerols but may decrease plasma HDL levels.

All of the drugs may have hepatic effects that lead to increases in certain liver enzymes such as alkaline phosphatase and glutamic-aspartate transaminase in plasma. Therefore, all patients need to be carefully monitored. The very long-term effects (over a period of 30–40 years) of these drugs have not been determined. Cholestryamine and nicotinic acid have been in use for many years, whereas Mevacor has been available as a prescription drug only since late 1987. Neomycin (not listed) has not found widespread use because of its potential harmful side effects on the kidneys and ears.

The costs of these lipid-lowering drugs vary widely. At present, the most expensive drug is Mevacor and the least expensive is niacin. Colestid and Lopid are less expensive than Mevacor but more expensive than niacin. The approximate cost of one 20-mg pill of Mevacor is $1.50, one 300-mg Lopid pill or 5 g of Colestid (or cholestyramine) costs 50 cents, and one 500-mg niacin pill costs 3 cents. Slow-release niacin pills are now available and cost more than regular niacin. Thus, the monthly costs of these drugs can vary from as little as $2.00 for niacin (1 g/day dose) to as high as $90 for Mevacor (two 20-mg pills per day).

8.2. COMBINED DRUG THERAPY

More dramatic effects in lowering plasma cholesterol levels in humans have been achieved by using a combination of two drugs. The combinations that have been successful are Colestid plus niacin, Colestid plus Lopid, or Colestid plus Mevacor. A combination of drugs is more effective than either drug alone because the drugs act at different points in cholesterol and lipoprotein metabolism. Indeed, whereas each drug alone may give a reduction of plasma cholesterol about 10–30%, the combination of drugs lowers plasma cholesterol by 30–60%. The actual amount of reduction depends on the dose of each drug and on patient compliance in taking the drugs. Diets low in cholesterol and saturated fats augment the effectiveness of the drugs. On the other hand, patients who feel that they can indulge in food because they are on medication can lessen the effectiveness of these lipid-lowering drugs.

The CLAS (cholesterol-lowering atherosclerosis) study was a randomized, placebo-controlled, angiographic trial testing Colestid (30 g/day) and niacin (3–12 g/day) therapy in 162 nonsmoking men ages 40–59 years with previous coronary bypass surgery. During the 2 years of treatment there was a 26% reduction in total plasma cholesterol, a 43% reduction in LDL-cholesterol, and a simultaneous elevation of HDL-cholesterol. This resulted in a significant reduction in the average number of lesions per subject that progressed and the percentage of subjects with new atheroma formation in native coronary arteries. Also, the percentage of subjects with new lesions or any adverse change in coronary artery bypass graphs was significantly reduced. Deterioration in overall coronary status was significantly less in drug-treated subjects. Regression of atherosclerosis, as indicated by perceptible improvement in overall coronary status, occurred in 16.2% of Colestid–niacin-treated subjects versus 2.4% in the placebo group (Blankenhorn et al., 1987). This study shows that plaque regression can occur and gives hope that further studies will provide more information on the factors that lead to plaque regression.

The actions of the drugs in lowering plasma cholesterol levels are depicted schematically in Fig. 8-2. The diagram on the left shows the normal situation in liver before drug therapy. The liver takes up LDL from the plasma and degrades them with lysosomal enzymes, releasing free cholesterol, amino acids, and other products. The liver also synthesizes cholesterol from acetate. The important intermediate is HMG-CoA, which is converted to mevalonic acid by the rate-limiting enzyme HMGR.

Excess cholesterol is converted to bile acids, the major part of which are reabsorbed via the enterohepatic cycle. As the cholesterol level in the cell rises, the liver decreases the synthesis of cholesterol and of LDL receptors. Fig. 8-2 (center) shows the effect of blocking bile acid reabsorption with Colestid which causes bile acid depletion by binding and sequestering the bile acids. This induces the liver to convert more cholesterol to bile acids, which lowers the cholesterol level in the cell and signals the cell to make more LDL receptors and to synthesize more cholesterol from acetate. As a result there is a modest lowering of plasma LDL. Fig. 8-2 (right) shows the effect of using two drugs in combination. One HMGR inhibitory drug (Mevacor) decreases the synthesis of cholesterol in the liver. The second drug (Colestid) prevents the reabsorption of bile acids. The combined effect of these drugs is to induce the liver to make more LDL receptors which increases the uptake of plasma LDL and thereby lowers plasma LDL. This combined effect leads to a dramatic lowering of plasma LDL and thus a lower plasma cholesterol. Normally two drugs are used in combina-

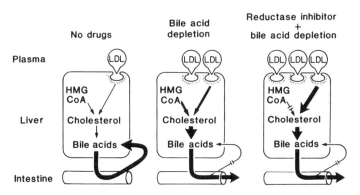

Figure 8-2. How drugs lower plasma lipids. (From Brown and Goldstein, 1986. Reproduced with permission.)

tion, but there is no reason why three drugs cannot be used together provided that their doses are carefully regulated so as to prevent toxicity. Lopid or niacin can be used as a third drug. It is noteworthy that the drug actions described above apply to heterozygous type 2 hyperlipoproteinemia (familial hypercholesterolemia) since people with this condition have only one defective gene for the LDL receptor and their cholesterol levels can be reduced from high values of 300–400 mg/dl to values of 150–200 mg/dl. However, people who are homozygous for type 2 hyperlipoproteinemia, have cholesterol levels of 800–1200 mg/dl. Since they have inherited two defective genes for the LDL receptor, drug therapy is not effective for even with drug therapy these individuals may have cholesterol levels of 400–800 mg/dl or higher. These patients have very little or no LDL receptors and are refractory to drug therapy.

Persons with type 3 and 4 hyperlipoproteinemia can be treated effectively with drug therapy, since they have normal LDL receptors and can reduce their elevated plasma cholesterol levels from high values of 280–350 mg/dl to a normal range of 150–200 mg/dl, especially when drug therapy is administered in conjunction with dietary management.

8.3. ANTITHROMBIC DRUGS

A thrombus is a plug formed by blood proteins and platelets. When red blood cells are trapped within the thrombus, a blood clot is formed. Thrombosis represents a series of events involving the activation of components of the hemostatic mechanism: the vessel wall, platelets, coagulation cascade, the fibrinolytic system, prostaglandins, growth factors, and possibly other agents.

Thrombi can occur in arteries, veins, capillaries, or chambers of the heart. Thrombogenesis involves endothelial injury, platelet adherence, platelet aggregation and release of thrombogenic factors, thrombin generation, and fibrin formation. Thrombus dissolution involves plasmin generation and fibrinolysis (Fig. 8-3).

8.3.1. Endothelial Injury

Factors that can injure the endothelium of arteries and cause thrombosis include hyperlipidemia (high LDL), hypertension, smoking, chemical factors, immunologic factors, infection, and mechanical injury. High blood pressure may lead to high shear forces caused by vortex flow of blood. The high shear forces can damage the endothelial cells or platelets and red blood cells as they impinge on the endothelium. Circulating immune complexes and antibodies directed against cell surface antigens or by sensitized lymphocytes can injure endothelial

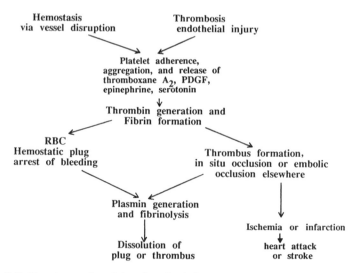

Figure 8-3. Thrombogenesis and thrombus dissolution. RBC, red blood cells. (From Harlan and Harker, 1983. Modified with permission.)

cells. Cigarette smoke may contain allergens that induce an immunoglobulin E response, which may injure endothelial cells.

Chemical injury to endothelial cells has been related to high LDL levels, to elevated levels of β-VLDL, and to chemicals such as homocystine, bile salts, radiologic contrast dyes and chemotherapeutic drugs. Anoxia can cause injury to the adventitia of arteries which can then damage endothelial cells. Carbon monoxide caused by cigarette smoking may lead to anoxia in arteries and damage the cells of the intima, media, and adventia.

8.3.2. Platelet Adherence and Aggregation

Vascular injury followed by endothelial denudation exposes the subendothelium collagen, elastin, and proteoglycans to circulating platelets. One hypothesis states that collagen binds to specific membrane proteins on the platelet surface, leading to platelet adherence. One such membrane glycoprotein is called GPIb. Plasma vonWillebrand factor (vWF) is also involved. vWF is synthesized by endothelial cells on megakaryocytes and also occurs in the α-granules of platelets. vWF binds to subendothelial collagen. GPIb may act as a receptor for vWF. Patients who lack GPIb (Bernard-Soulier syndrome) or vWF (von-Willebrand's disease), have platelets which do not adhere well. Pigs that lack vWF or have severe thrombocytopenia have been shown to be very resistant to atherosclerosis induced by injury with balloon catherization or by injection of homocystine.

Platelet aggregation follows platelet binding to subendothelial collagen (and possible to subendothelial proteoglycans). Aggregation involves a drastic change in the shape of platelets and exposure of new binding sites for fibrinogen and for other clotting factors such as prothrombin. The platelet membrane has the acidic phospholipids phosphatidylserine (PS) and phosphatidylethanolamine (PE) localized on the inner half of the lipid bilayer. Consequently, these phospholipids of normal platelets are not exposed to the plasma. The same holds true for red cells. However, when platelet aggregation occurs, the lipid bilayer of the membrane is damaged and exposes PS and PE. These lipids serve as sites for binding Ca^{2+} and activating prothrombin. Prothrombin is converted to thrombin, which in turn converts fibrinogen to fibrin. The fibrin forms a network that binds platelets, especially platelets that have been aggregated and have binding sites for fibrin. This trapping of platelets on the fibrin network leads to a thrombus. Red blood cells can also be trapped and form a clot.

The aggregated platelets undergo a reacting leading to the release of ADP, serotonin, epinephrine, Ca^{2+}, PDGF, β-thromboglobulin (β-TG) and platelet factor 4 (PF4). Aggregation also activates phospholipase A_2 in the platelet membrane. Phospholipase A_2 hydrolyzes membrane phospholipids to release arachidonic acid, which is converted to TXA_2 by several enzyme systems, the first being prostaglandin cyclooxygenase. TXA_2 and ADP are very potent platelet-aggregating agents that serve to amplify the aggregation process (Fig. 8-4).

8.3.3. Thrombin and Fibrin Formation

Growth and stabilization of the thrombus plug involves the formation of thrombin at the site of vascular injury. Thrombin is generated from prothrombin when tissue thromboplastin from injury cells activates the extrinsic coagulation cascade but also is produced by the intrinsic coagulation pathway. Injury or aggregated platelets contribute to thrombin formation by providing membrane phospholipids (PE and PS) and receptors for coagulation factors XI, V, and VIII. These facilitate the conversion of factor X to factor Xa. Factors Xa and V act to convert prothrombin to thrombin. Within seconds after blood is exposed to the

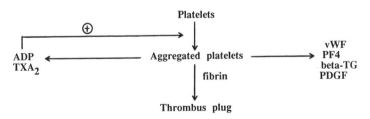

Figure 8-4. Events in platelet aggregation. ⊕, Stimulation. (From Harlan and Harker, 1983. Modified with permission.)

denuded intima, sufficient thrombin is formed to stimulate platelet α-granule release. Thrombin binds to specific receptors on the damaged platelet surface. The binding releases ADP and also activates phospholipase A_2 releasing arachidonic acid, which is converted to TXA_2. ADP and TXA_2 further enhance platelet aggregation. Thrombin converts fibrinogen to fibrin, forming a network that stabilizes the plug.

8.3.4. Plasmin Formation and Fibrinolysis

Fibrinolysis (dissolution of the thrombus plug) occurs within the thrombus. When fibrin forms, it binds plasminogen, which in turn binds to and activates tissue plasminogen activator (TPA). Thrombin, together with thrombomodulin, activates a specific protein C that increases the release of TPA. The activation of plasminogen which is bound to fibrin leads to the release of plasmin. The generation of plasmin within the thrombus protects plasmin from inactivation by plasma α2-antiplasmin. Plasmin digests fibrin and causes dissolution of the thrombus, whih now exposes plasmin to α2-antiplasmin and terminates the action of plasmin.

An overall scheme showing the pathways for thrombus formation and dissolution is shown in Fig. 8-5.

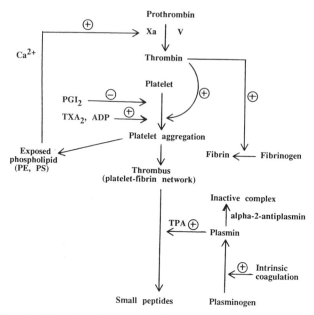

Figure 8-5. Thrombus formation and dissolution. ⊕, Stimulation; ⊖, inhibition. (From Harlan and Harker, 1983. Modified with permission.)

8.4. FIBRINOLYTIC DRUGS

As mentioned above, dissolution of a thrombus plug is mediated by specific proteolytic enzymes. Plasmin is a serine protease that cleaves Arg–X and Lys–X peptide bonds in fibrin to yield small peptide fragments (X = some other amino acid). Once plasmin has done its job and diffuses from the dissolved plug, it is inactivated by a protease inhibitor, α2-antiplasmin.

Streptokinase is a protein excreted by group C β-hemolytic streptococci. It forms a complex with plasminogen and in the process converts plasminogen to plasmin. When streptokinase is infused into a patient, a sufficient amount must be used to overcome circulating antistreptococcal antibodies and plasma α2-antiplasmin. Since free plasmin generated by streptokinase will digest not only fibrin in the thrombus but also circulating fibrinogen and factors V and VIII, and since free plasmin does not distinguish between thrombus and normal circulating coagulation factors, fibrinolysis thereby can lead to serious bleeding problems.

TPA binds to fibrin and becomes activated specifically at the thrombus site, where it then converts plasminogen to plasmin. This produces plasmin locally at the thrombus and avoids the complications of streptokinase or urokinase. TPA is believed by some clinicians to be the fibrinolytic agent of choice in the treatment of coronary occlusions and can be given intravenously rather than with a catheter, as required for streptokinase. However, other clinicians maintain that streptokinase is more effective since it is 1/10 as costly ($200 versus $2200 per dose) as TPA and mortality studies show that it is as effective as TPA, even though TPA opens clots 65–75% of the time whereas streptokinase opens clots 40–50% of the time. Moreover, streptokinase is given intravenously over the span of 1 hr whereas at least 3 hr is required to administer TPA. Some clinicians will not use TPA after 4 hr of onset of chest pain because its advantage over streptokinase is lost at that point. Some preliminary evidence indicates that TPA may cause more bleeding in the brain, which may lead to brain damage. On the other hand, more patients are allergic to streptokinase. However, the percentage of such patients is very low (1.5%), and these patients do not die from the allergic reactions that streptokinase can cause. The Second International Study of Infarct Survival, involving about 17,000 patients showed that streptokinase or aspirin alone reduced mortality to myocardial infarction (MI) by about 25%, but when the two agents were given together they reduced mortality by more than 40%. Finally, one must consider that both drugs are most effective within 4 h of onset of chest pain, both can dissolve clots that have closed ulcers (causing the ulcers to start bleeding again), and both can exacerbate bleeding of healing clots after surgery. Use of buffered or slow-release aspirin can circumvent exacerbation of stomach ulcers. The incidence of strokes is quite low (about 0.5–1.0%) in patients who receive thrombolytic agents. This level is considered safe considering the life-saving benefits of the therapy.

A new thrombolytic agent called APSAC (anisolyated plasminogen streptokinase activator complex) has been made available that, like TPA, is supposed
to be clot specific. Newer forms of both TPA and streptokinase are being developed that hopefully will have less deleterious side effects and be more effective
than the current drugs used to treat MI.

8.5. MECHANISMS OF ACTION OF ANTITHROMBIC DRUGS

Because platelet aggregation plays a major role in atherogenesis, there has
been great interest in agents that inhibit platelet aggregation. These antithrombic
agents act at different sites in the complex process of thrombosis (Fig. 8-6).
The drugs that inhibit the release of arachidonic acid or its conversion to
TXA_2 include the nonsteroidal antiinflammatory agents and thromboxane synthetase inhibitors (dipyridamole) and cyclooxygenase inhibitors (aspirin, phe-
nylbutazone, and indomethacin). Inhibitors of the arachidonic pathway will not
affect other mechanisms of platelets in thrombus formation such as adherence of
platelets to collagen. Cyclooxygenase and thromboxane synthetase inhibitors
prevent recruitment of circulating platelets but do not inhibit binding of platelets
to the subendothelium.

Platelet aggregation is involved in thrombus formation and in setting in
motion the process of atherosclerosis. Heart attacks and strokes are caused by
thrombi in small arteries of the heart and brain respectively. Therefore, it stands
to reason that inhibition of thrombus formation by inhibiting platelet aggregation, would help prevent heart attacks and strokes. This effect can be obtained by
inhibiting the formation of agents produced in the body that enhance platelet
aggregation. Since TXA_2 (made by platelets when they are damaged) is a major
stimulator of platelet aggregation, then a drug that inhibits the production of
TXA_2 should be beneficial in preventing heart attacks and strokes.

The common drug that provides this protection is aspirin (acetylsalicylic
acid). Indeed, only one baby aspirin (60 mg) per day is required to inhibit
completely TXA_2 production by platelets. This finding has led to numerous
studies on the effectiveness of aspirin in protecting humans from heart attacks
and strokes. Initial studies yielded mixed results because of the different dosages
used (one to four aspirin tablets, equivalent to 325–1300 mg per day). A dose of
three aspirin tablets per day is commonly used to treat men who have had
ministrokes (transient ischemic attacks). This treatment provides a 50% reduction in stroke occurrence in men but has been less effective in women.

A British study found a 25% reduction in recurrent heart attacks among
patients taking one aspirin pill (325 mg) a day. In seven studies done in the
United States, Canada, and Europe, five indicated favorable results and three
gave equivocal results based on mortality rate and number of nonfatal MI. More

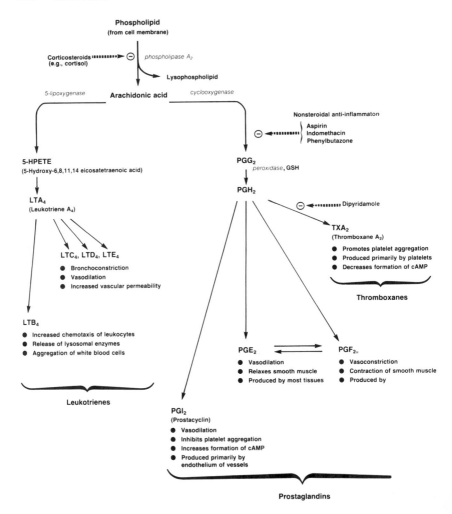

Figure 8-6. Biosynthesis and function of prostaglandins, thromboxanes, and leukotrienes and points of action of antithrombic drugs. (From Champe and Harvey, 1987. Reproduced with permission.)

importantly, however, a very recent study (Young et al., 1988) by Harvard University on 22,071 male physicians has shown that taking one aspirin every other day provides 40% protection against fatal heart attack. It is expected that similar protection is afforded women.

The dose of aspirin given is critical for the following reason. Two prostaglandins are involved in platelet aggregation. One, TXA_2, made in the platelet, is a very potent activator of platelet aggregation. The other, PGI_2, made in

arterial endothelial cells, is a very potent inhibitor of platelet aggregation. Therefore a delicate balance between these two prostaglandins must be maintained.
Aspirin acetylates and inhibits the synthesis of both thromboxane and prostacyclin since it inhibits irreversibly the enzyme cyclooxygenase, which
catalyzes the first reaction leading to the synthesis of all prostaglandins. However, platelets do not contain nuclei and cannot make new enzymes, whereas
arterial intima cells are nucleated and can synthesize new enzymes. Moreover,
platelets live for only 7–9 days. The key issue is use of the minimum amount of
aspirin that will effectively inhibit thromboxane synthesis in platelets but mini-
mally inhibit prostacyclin synthesis in arterial cells. Moreover, the effect on
platelets lasts for the life of the platelet, whereas the effect on arterial cells is
short, ending when the aspirin level falls and the arterial cells make new enzymes
that make more prostacyclin. The optimum dose of aspirin is clearly very important, and current studies indicate that about one baby aspirin (60 mg) per day is
sufficient to get this favorable balance between prostacyclin and thromboxane.
The major benefit from aspirin then, is prevention of platelet aggregation and
clots, which can obstruct arteries and cause heart attacks or strokes. Like any
drug, aspirin has side effects in some humans. In particular, some young children
may be more susceptible to Reye's syndrome when they take aspirin. Aspirin
also can aggravate symptoms associated with peptic ulcers. There is concern
among some researchers and physicians that the daily intake of aspirin may
increase the risk of stroke especially in the elderly population.

Sulfinpyrazone is a competitive reversible inhibitor of cyclooxygenase. Unlike the case with aspirin, doses of sulfinpyrazone that inhibit cyclooxygenase
neither prolong bleeding times in normal persons nor measurably affect in vitro
platelet aggregation. Thromboxane synthetase inhibitors such as dipyridamole
block the conversion of cyclic endoperoxides to TXA_2. These types of inhibitors
should be more effective than cyclooxygenase inhibitors, since they preferen-
tially inhibit TXA_2 without influencing PGI_2 synthesis. The long-term clinical
effects of thromboxane inhibitors remain to be determined.

Eicosapentenoic acid found in fish oils has been shown to inhibit the synthesis of arachidonic acid and also the conversion of endoperoxides to TXA_2.
Furthermore, this fatty acid is converted by platelets to TXA_3 which does not
enhance platelet aggregation and is converted by endothelial cells to PGI_3, which
inhibits platelet aggregation.

8.6. ARACHIDONIC ACID RELEASE INHIBITORS

Drugs that inhibit the membrane phospholipase A_2 reduce the release of
arachidonic acid from membrane phospholipids and reduce TXA_2 synthesis in
platelets. These drugs include corticosteroids (Fig. 8-6).

8.7. DRUGS THAT INHIBIT cAMP FORMATION IN PLATELETS

When PGI_2 binds to its specific membrane receptor, it activates adenyl cyclase, an enzyme that converts ATP to cAMP. Elevated levels of cAMP activate a protein kinase that phosphorylates and inactivates platelet myosin light-chain kinase and thereby inhibits platelet contraction and aggregation. PGI_2 exerts its antiaggregation effect on platelets by this mechanism. TXA_2, on the other hand, inhibits adenyl cyclase and decreases cAMP levels, thereby increasing platelet aggregation.

Dipyridamole (Persantin) also acts by inhibiting the phosphodiesterase that degrades cAMP to 5'-AMP, thus increasing the levels of cAMP. Since PGI_2 increases cAMP production and dipyridamole decreases cAMP breakdown, a synergistic action prevails when these drugs are used in combination. Unlike aspirin and sulfinpyrazone, dipyridamole reduces platelet adherence to subendothelial collagen. Dipyridamole also inhibits the conversion of PGH_2 to TXA_2 in platelets.

8.8. MEMBRANE-ACTIVE DRUGS

In addition to corticosteroids, which act on membrane-bound phospholipase A_2, some drugs, such as ticlopidine appear to change the sensitivity of the platelet membrane by altering its fluidity or modifying membrane receptors. Ticlopidine is a strong inhibitor of platelet function that prevents primary aggregation of platelets induced by nearly all agonists. It may act in part by inhibiting vWF and fibrinogen interactions with GPI6.

8.9. ANTICOAGULANTS AND OTHER DRUGS

The β-adrenergic antagonist propranolol (Inderal) decreases the enhanced platelet sensitivity to ADP-induced aggregation. Organic nitrate vasodilators also have been reported to inhibit platelet aggregation. However, these drugs have other, more important effects on the heart that make them very useful in treating angina pectoris and myocardial ischemia.

Heparin and warfarin interfere with thrombus formation by preventing the action of or synthesis of thrombin. Heparin is an acidic sulfonated mucopolysaccharide containing repeating units of iduronic acid and sulfated N-acetylglucosamine. The anticoagulant activity of heparin is related to a specific tetrasaccharide sequence of two iduronic acid residues. The tetrasaccharide binds to lysine residues on antithrombin III (AT-III) and markedly reduces the activity of a specific positively charged arginine group of AT-III. This arginine group binds to the negatively charged active site of serine proteases (thrombin, factors IXa, Xa, XIa, and VIIa, and plasmin) and inhibits these proteases.

Warfarin (dicumarol) is a vitamin K antogonist, since it is structurally similar to vitamin K. Warfarin inhibits the action of vitamin K, which stimulates the posttranslational carboxylation in liver of blood coagulation factors pro-thrombin, VII, IX, and X. This inhibition of carboxylation of specific glutamic residues on these coagulation factors prevents the formation of a highly nega-tively charged site that is necessary for Ca^{2+} binding. The Ca^{2+} binding is required for these factors to interact with acidic phospholipids on the platelet membrane. Ca^{2+} acts as a bridge between the coagulation factors and the platelet membrane.

8.10. ANTITHROMBIC DRUGS FOR TREATMENT OF CHD

Transmural MI is a thrombotic event. In contrast to patients with MI, most patients with sudden cardiac death do not have cardiac necrosis or coronary artery thrombi. Large thrombi do not appear to play a role in sudden cardiac death or in angina pectoris but do play a major role in MI. Thus, the major trials of antithrombic therapy have focused on MI. The first episode of an MI shows the failure of the antithrombic regimen in primary prevention trials, and reoccur-rence of an MI serves as a valid endpoint in secondary prevention trials. How-ever, some patients with transmural infarcts may die suddenly and cannot be distinguished from patients with unheralded sudden death.

Each year in the United States there are about 1 million heart attacks (includes acute MI and sudden cardiac death). Nearly 200,000 victims die before admission to a hospital, and another 200,000 die in the first month, most in the first 24 hr after the attack. Thus, the cumulative mortality in the first month is close to 40%. During the next 6 months, the cumulative mortality increases another 7–8% so that by one year, additional 10% of patients have died. This means that of the 400,000–500,000 patients who survive the first month about 40,000–50,000 die during the first year after an MI. After the first year, the reinfarction rate and death rate stabilize to 3–5% per year. This rate is similar to that of patients with symptomatic coronary artery disease but is significantly greater than that of a randomly selected age-matched population without overt coronary artery disease that has a 1% yearly incidence.

Intervention is best done before an acute MI. However, a primary preven-tion trial in an unselected population requires many thousand patient-years of study to detect a 30% reduction in mortality or infarction if the frequency is 1% per year. This involves a large cost and thousands of patients and is complicated by the safety of the drug and patient compliance. For this reason, secondary prevention trials are more practical, even though the patients are late in their diseases. Moreover, fewer patients are required for the study.

8.10.1. Anticoagulant Therapy for MI

In the late 1940s and 1950s a number of trials were performed to evaluate short-term anticoagulation therapy on mortality and reinfarction in patients with acute MI. Many of these studies were not randomized and used historical, alternate, or simultaneously collected controls. Methodologic problems precluded any definite conclusions. More recently, six randomized, controlled, prospective studies have been carried out by the Medical Research Council, Bronx Municipal Hospital, and Veterans Administration Hospital. The mortality and reinfarction rates were not significantly different from those of controls except for women in the Bronx Municipal Hospital study. The failure of anticoagulants to reduce mortality is not surprising, since most early deaths following MI are due to dysrhythmias and pump failure.

Studies in the 1980s attempted to generate new enthusiasm for the short-term use of anticoagulants in MI. In these retrospective studies from Israel and Maryland, a two- to threefold lower mortality was reported among treated patients. These studies were criticized on the basis of patient selection bias. Reexamination of the properly designed randomized trials showed that when case/fatality ratios were pooled, the small differences in mortality in each study became statistically significant. The overall reduction in mortality was 21%.

8.10.2. Long-Term Anticoagulant Therapy for MI

Most of the early studies done between 1950 and 1960 were suspect because of the small number of patients used. In 1970, an International Anticoagulant Review Group tried to overcome this problem by pooling data from nine controlled trials involving 2205 men and 282 women. The review group concluded that mortality was reduced 20% in treated men, especially those with prolonged angina or previous MI.

In 1980, the Sixty Plus Reinfarction Study from the Netherlands reviewed the issue of the value of long-term anticoagulant therapy. In six centers, ambulant patients over 60 years old were studied. All patients were receiving therapy following documented MI that occurred at least 6 months earlier. A randomized double-blind study of 439 patients receiving placebo and 439 patients receiving anticoagulant therapy showed that numbers of total deaths and sudden deaths were not significantly different in the two groups. However, there was a dramatic and clear-cut 55% reduction in the incidence of fatal and nonfatal MI in the treated group.

8.10.3. Platelet Inhibitors and Acute MI

In the 1960s studies were done using dipyridamole, sulfinpyrazone, and aspirin. Over the past 10 years there have been eight randomized controlled studies of these drugs in the secondary prevention of MI, involving more than 10,000 patients at a cost in excess of $30 million. Unfortunately, the results are not conclusive in part because of the difference in dose of each drug used and duration of the trials. Five of the studies showed a trend in favor of aspirin when total number of deaths was the endpoint. Four trials showed a 34–50% reduction in mortality or in coronary incidence. Except for one trial (Aspirin Myocardial Infarction Study involving 4,524 patients), the trend was in favor of using platelet-modifying agents.

As mentioned above, hovever, new studies at Harvard University on 22,071 healthy U.S. male physicians showed that taking one 325 mg aspirin tablet every other day gave 40% protection against first time-heart attacks. This 10-year study began in 1980. The results were so dramatic that the preliminary findings were made available in 1987. It must be kept in mind that, in theory, the proper use of these drugs, especially the appropriate dose to obtain a favorable PGI_2/TXA_2 ratio, should have long-term beneficial effects based on the premise that platelet aggregation is intimately involved in atherogenesis. Too much aspirin can inhibit PGI_2 synthesis, and this may enhance clotting. Studies indicate that 60–80 mg of aspirin per day is sufficient to inhibit platelet aggregation without having a deleterious effect on the synthesis of PGI_2 by endothelial cells.

A prospective randomized double-blind trial by Chesebro et al. (1984) showed that the combined use of dipyridamole (Persantin) and aspirin markedly increased graft patency at 1 month following surgery without increasing postoperative bleeding.

Although antithrombic and fibrinolytic drugs have found widespread use for the treatment of MI and CHD, it is expected that in the near future newer drugs and the more effective use of drugs in combination will prove to be more effective.

REFERENCES

Aspirin and blood clotting, 1988, *N. Engl. J. Med.*, 309:396.

Atmeh, R. F., Stewart, J. M., Boag, D. E., Packard, C. J., Lorimer, A. R., and Shepherd, J., 1983, The hypolipidemic action of probucol: a study of its effects on high and low density lipoproteins, *J. Lipid Res.*, 24:588.

Been, M., deBono, D.P., Boulton, A. L., Hillis, W. S., and Hornung, R., 1985, Coronary thrombolysis with anisolyated plasminogen-streptokinase complex, *Brit. Heart J.*, 53:253.

Benditt, E. P., and Schwarz, S. M., 1984, Atherosclerosis: what can we learn from studies in human tissues, *Lab. Invest.*, 50:3.

Blank, D. W., Hoeg, J. M., Kroll, M. H., and Ruddel, M. E., 1986, The method of determination must be considered in interpreting blood cholesterol levels, *J. Am. Med. Assoc.*, 256:2867.

Blankenhorn, D. H., Nessim, S. A., Johnson, R. L., Sanmarco, M. E., Azen, S. P., and Cashin-Hemphill, L., 1987, Beneficial effects of combined colestipol-niacin therapy on coronary atherosclerosis and coronary venous bypass grafts, *J. Am. Med. Assoc.*, 257:3233.

Brown, M. S., and Goldstein, J. L., 1986, A receptor-mediated pathway for cholesterol homeostasis, *Science*, 232:34.

Canner, P. L., Berge, K. G., Wenger, N. K., Stamler, J., Friedman, L., Prineas, R. J., and Friedewald, W., 1986, Fifteen year mortality in coronary drug project patients:long-term benefit with niacin, *J. Am. Coll. Cardiol.*, 8:1245.

Castelli, W. P., Garrison, R. J., Wilson, P. W. F., Abbott, R. D., Kalousdian, S., and Kannel, W. B., 1986, Incidence of coronary heart disease and lipoprotein cholesterol levels, the Framingham study, *J. Am. Med. Assoc.*, 256:2835.

Champe, P. C., and Harvey, R. A., 1987, *Lippincott's Illustrated Reviews: Biochmistry*, J. B. Lippincott Co., Philadelphia.

Chesebro, J. H., Fuster, V., Elveback, L. R., Clements, I. P., Smith, H. C., Holmes, D. R., Jr., Bardsley, W. T., Pluth, J. R., Wallace, R. B., Puga, F. J., Orszulak, T. A., Piehler, J. M., Danielson, G. K., Schaff, H. V., and Frye, R. L., 1984, Effect of dipyridamole and aspirin on late vein-graft patency after coronary bypass operations, *N. Engl. J. Med.*, 310:209.

Consensus conference, 1985, Lowering blood cholesterol to prevent heart disease, *J. Am. Med. Assoc.*, 253:2080.

East, C., Grundy, S. M., and Bilheimer, D. W., 1986, Normal cholesterol levels with lovastatin (mevinolin) therapy in a child with homozygous familial hypercholesterolemia following liver transplantation, *J. Am. Med. Assoc.*, 256:2843.

Effect of prostaglandin I3 formation in vivo from dietary eicosapentaenoic acid, 1984, *Nutr. Rev.*, 42:317.

Etingin, O. R., Weksler, B. B., and Hajjar, D. P., 1986, Cholesterol metabolism is altered by hydrolytic metabolites of prostacyclin in arterial smooth muscle cells, *J. Lipid Res.*, 27:530.

Expert Panel, 1988, Report of the national cholesterol education program expert panel on detection, evaluation, and treatment of high blood cholesterol in adults, *Arch. Inter. Med.*, 148:36.

Frick, M. H., Elo, O., Haapa, K., Heinonen, O. P., Heinsalmi, P., Helo, P., Huttunen, J. K., Kaitaniemi, P., Koskinen, P., Manninen, V., Maenpaa, H., Malkonen, M., Manittari, M., Norola, S., Pasternack, A., Pikkarainen, J., Romo, M., Sjoblom, T., and Nikkila, E. A., 1987, Helsinki heart study: primary-prevention trial with gemfibrozil in middle-aged men with dyslipidemia, *N. Engl. J. Med.*, 317:1237.

Goldberg, R. J., Gore, J. M., Dalen J. E., and Alpert, J. S., 1986, Long-term anticoagulant therapy after acute myocardial infarction, *Am. Heart J.*, 109:616.

Goldbourt, U., Holtzman, E., and Neufeld, H. N., 1985, Total and high density lipoprotein cholesterol in the serum and risk of mortality: evidence of a threshold effect, *Brit. Med. J.*, 290:1239.

Gotto, A. M., Robertson, A. L., Epstein, S. E., DeBakey, M. D., and McCollum, C. H., 1980, in: *Atherosclerosis*, (H.L. Gross, ed.), Sections I and II, The Upjohn Co., Kalamazoo, Michigan.

Grundy, S. M., 1986, Cholesterol and coronary heart disease. A new era, *J. Am. Med. Assoc.*, 256:2849.

Grundy, S. M., and Vega, G. L., 1985, Influence of mevinolin on metabolism of low density lipoproteins in primary moderate hypercholesterolemia, *J. Lipid Res.*, 26:1464.

Harlan, J. M., and Harker, A., 1983, *Thrombosis and Coronary Artery Disease*, The Upjohn Co., Kalamazoo, Michigan.

Hegele, R. A., Huang, R. A., Herbert, P. N., Blum, C. B., Buring, J. E., Hennekens, C. H., and

Breslow, J. L., 1986, Apolipoprotein B-gene DNA polymorphisms associated with myocardial infarction, *N. Engl. J. Med.*, 315:1509.

Huff, M. W., Telford, D. E., Woodcroft, K., and Strong, W. P., 1985, Mevinolin and cholestryamine inhibit the direct synthesis of low density lipoprotein apolipoprotein B in miniature pigs, *J. Lipid Res.*, 26:1175.

Huddleston, C. B., Hammon, J. W., Wareing, T. H., Lupinetti, F. M., Clanton, J. A., Collins, J. C., and Bender, H. W., 1985, Ameloriation of the deleterious effects of platelets activated during cardiopulmonary bypass: comparison of a thromboxane synthetase inhibitor and a prostacyclin analogue, *J. Thorac, Cardiovasc. Surg*, 89:190.

Illingworth, D. R., Phillipson, B. E., Rapp, J. H., and Connor, W. E., 1981, Colestipol plus nicotinic acid in treatment of heterozygous familial hypercholesterolaemia, *Lancet*, i:296.

Jonasson, L., Bondjers, G., and Hansson, G. K., 1987, Lipoprotein lipase in atherosclerosis: its presence in smooth muscle cells and absence from macrophages, *J. Lipid Res.*, 28:437.

Kane, J. P., Malloy, M. J., Tun, P., Phillips, N. R., Freedman, D. D., Williams, M. L., Rowe, J. S., and Havel, R. J., 1981, Normalization of low-density-lipoprotein levels in heterozygous familial hypercholesterolemia with a combined drug regimen, *N. Engl. J. Med.*, 304:251.

Kinosian, B. P., and Eisenberg, J. M., 1988, Cutting into cholesterol: cost-effective alternatives for treating hypercholesterolemia, *J. Am. Med. Assoc.*, 259:2249.

Krause, B. R., and Newton, R. S., 1985, Apolipoprotein changes associated with the plasma lipid-regulating activity of gemfibrozil in cholesterol-fed rats, *J. Lipid Res.*, 26:940.

Kromhout, D., Bosscheiter, E. B., and De Lezenne Coulander, C., 1985, The inverse relation between fish consumption and 20-year mortality from coronary heart disease, *N. Engl. J. Med.*, 312:1205.

Kuo, P. T., Kostis, J. B., Moreyra, A. E., and Hayes, J. A., 1981, Familial type II hyperlipoproteinemia with coronary heart disease. Effect of diet-colestipol-nicotinic acid treatment, *Chest*, 79:286.

Lowering blood cholesterol to prevent heart disease, 1985, *Nutr. Rev.*, 43:283.

Molecular biology of coronary heart disease, 1987, *Nutr. Rev.*, 45:108.

Naruszewicz, M., Carew, T. E., Pittman, R. C., Witztum, J. L., and Steinberg, D., 1984, A novel mechanism by which probucol lowers low density lipoprotein levels demonstrated in the LDL receptor-deficient rabbit, *J. Lipid Res.*, 251:1206.

Packard, C. J., and Shepherd, J., 1982, The hepatobiliary axis and lipoprotein metabolism: effects of bile acid sequestrants and ileal bypass surgery, *J. Lipid Res.*, 23:1081.

Packard, C. J., Clegg, R. J., Dominiczak, M. H., Lorimer, A. R., and Shepard, J., 1986, Effects of bezafibrate on apolipoprotein B metabolism in Type III hyperlipoproteinemic subjects, *J. Lipid Res.*, 27:930.

Pedersen, A. K., and Fitzgerald, G. A., 1984, Dose-related kinetics of aspirin: presystemic acetylation of platelet cyclooxygenase, *N. Engl. J. Med.*, 311:1206.

Rapp, J. H., Connor, E. E., Lin, D. S., Inahara, T., and Porter, J. M., 1983, Lipids of human atherosclerotic plaques and xanthomas: clues to the mechanism of plaque formation, *J. Lipid Res.*, 24:1329.

Regulation of plasma cholesterol by compactin and mevinolin, *Nutr. Rev.* 43, 266 (1985).

The lipid research clinics coronary primary prevention trial results. I. Reduction in incidence of coronary heart disease, 1984, *J. Am. Med. Assoc.*, 251:351.

Roberts, A. B., Lees, A. M., Strauss, H. W., Fallon, J. T., Taveras, J., and Kopiwoda, S., 1983, Selective accumulation of low density lipoproteins on damaged arterial wall, *J. Lipid Res.*, 24:1160.

Ross, R., and Glomset, J. A., 1976, The pathogenesis of atherosclerosis, *N. Engl. J. Med.*, 295:369, 420.

Rudel,L. L., Parks, J. S., Johnson, F. L., and Babiak, J., 1986, Low density lipoproteins in atherosclerosis, *J. Lipid Res.*, 27:465.

Scanu, A. M., Wissler, R. W., and Getz, G. S. (eds.), 1979, *The Biochemistry of Atherosclerosis*, Marcel Dekker, New York.

Shepherd, J., Packard, C. J., Bicker, S., Veitch Lawrie, T. D., and Gemmel Morgan, H., 1980, Cholestyramine promotes receptor-mediated low-density-lipoprotein catabolism, *N. Engl. J. Med.*, 301:1219.

Sherwin, R., and Chir, B., 1984, Recent clinical trials on the preventability of coronary heart disease, *Clin. Nutr.*, 3:128.

Sherwin, R., Chir, B., and Cutler, J. A., 1984, Antihypertensive agents on lipids, lipoproteins, and coronary heart disease, *Clin. Nutr.*, 3:131.

Stamler, J., Wentworth, D., and Neaton, J. D., 1986, MRFIT Research Group, Is relationship between serum cholesterol and risk of premature death from coronary heart disease continuous and graded? Findings in 356,222 primary screenes of the MRFIT, *J. Am. Med. Assoc.*, 256:2823.

Starzi, T. E., Bilheimer, D. W., Bahnson, H. T., Shaw, B. W., Jr., Hardesty, R. L., Griffith, B. P., Iwatsuki, S., Zitelli, B. J., Gartner, J. C., Jr., Malatack, J. J., and Urbach, A. H., 1984, Heart-liver transplantation in a patient with familial hypercholesterolemia, *Lancet*, i:1382.

Strong, J. P., 1986, Coronary atherosclerosis in soldiers. A clue to the natural history of athe-rosclerosis in the young, *J. Am. Med. Assoc.*, 256:2863.

The lovastatin study group, II, Therapeutic response to lovastatin (mevinolin) and nonfamilial hypercholesterolemia, 1986, *J. Am. Med. Assoc.*, 256:2829.

Wynder, E. L., Field, F., and Haley, N. J., 1986, Population screening for cholesterol determination, a pilot study, *J. Am. Med. Assoc.*, 256:2839.

Yamamoto, A., Matsuzawa, Y., Kishino, B., Hayashi, R., Hirobe, K., and Kikkawa, T., 1983, Effects of probucol on homozygous cases of familial hypercholesterolemia, *Atherosclerosis*, 48:157.

Young, F. E., Nightingale, S. L., and Temple, R. A., 1988, The preliminary report of the findings of the aspirin component of the ongoing physician's health study, *J. Amer. Med. Assoc.*, 259:3158.

Chapter 9

LUNG SURFACTANT DEFICIENCY
Respiratory Distress Syndrome of the Newborn

9.1. EARLY HISTORY

Respiratory distress syndrome (RDS) in infants was first described in 1959 by Avery and Mead who showed it results from a deficiency of surfactant, a phospholipid–protein complex, in the lung. RDS, also called hyaline membrane disease, is most prevalent in premature infants. This lung abnormality has interested many scientists and clinicians and has led to a greater understanding of the biochemistry of the lung and to the elucidation of the components of lung surfactant. To understand the function of lung surfactant, it is appropriate to consider first the overall gross anatomy of the lung (Fig. 9-1).

The alveoli greatly increase the surface area of the lung and allow the efficient exchange of carbon dioxide and oxygen. To prevent collapse of the alveoli during expiration, the lungs secrete a surfactant that lowers the surface tension of the alveoli. The complex nature of surfactant, which contains a variety of phospholipids and proteins, allows the surface tension of the alveoli to change during expiration and inspiration.

The key experiment providing evidence for an alveolar surface active material was published by von Neergaard in 1929. Later experiments showed that the pressure required to expand an air-filled lung is greater than that needed to expand a saline liquid-filled lung.

The pressure–volume relationships in the excised lungs of a cat are shown in Fig. 9-2. The pressure was held constant at each point, and the volume was allowed to come into equilibrium at that pressure. An important factor affecting the compliance of the lung is the surface tension. The magnitude of this component at various lung volumes can be measured by removing the lungs from the body and distending them alternatively with saline and with air while measuring

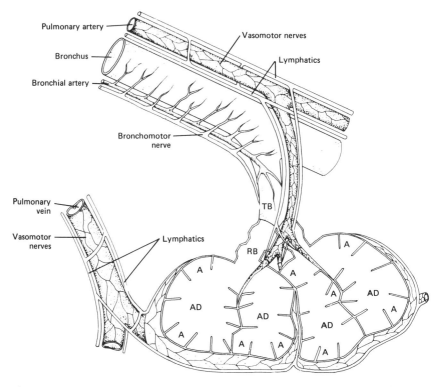

Figure 9-1. Structure of the human lung. A, anatomic alveolus; AD, alveolar duct; RB, respiratory bronchiole; TB, terminal bronchiole. (From Staub, 1970. Reproduced with permission.)

the intrapulmonary pressure. Because saline reduces the surface tension to nearly zero, the pressure–volume curve obtained with saline measures the tissue elasticity, while the curve obtained with air measures both components. The difference between the two curves is the elasticity due to surface tension. It is reduced as the volume of the lung is made smaller. The additional component of pressure is required to counter the inward-acting forces resulting from the alveolar air–liquid interface, similar to the situation in a soap bubble at the end of a tube. Unless there is a transmural pressure acting in an outward direction across the inner surface of the bubble, the inward-acting surface forces tend to reduce the bubble so as to attain a minimum surface area.

The Young-Laplace equation provides a quantitative measure of the pressure needed to stabilize the bubble: $P = 2q/r$, where P is pressure, q is surface tension, and r is radius. As surface tension increases or radius decreases, the transmural pressure must increase. In transposing this model to the lung, the transmural pressure acting across the alveolus is the transpulmonary pressure.

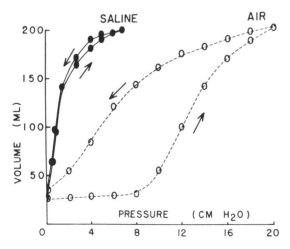

Figure 9-2. Pressure–volume curves on inflating and deflating an excised cat lung. The pressure was held constant at each point and the volume was allowed to come into equilibrium at that pressure. (From King, 1978. Reproduced with permission.)

However, the magnitude of the surface tension is not the only factor affecting alveolar stability. The surface tension is variable and changes with surface area. If alveoli were lined with a surfactant of constant surface tension at all volumes, then the alveoli would be unstable no matter what the magnitude of the surface tension (other than zero). If the alveoli were lined with a surface film (surfactant) so that surface tension decreased as the alveolar volume decreased, the alveoli would be stabilized, since the transpulmonary pressure would decrease as the volume decreased.

The way in which surface tension varies with surface area is known as surface elastance, (S), defined as $Adq/dA = dq/dlnA$, where q is surface tension and A is surface area. The alveolar stability is favored by low surface tension, high surface elastance, and large transpulmonary pressure as shown by the equation $q_{min} = 8q - 4S/3P$, where q_{min} is the minimum radius of an alveolus that can be stable at given values of surface tension (q), surface elastance (S) and transpulmonary pressure (P) (Clements, 1970). This equation reveals that alveoli with radii greater than 30–40 μm would be stable. The lung cannot be described simply as the equations presented above indicate, since other factors are involved: (1) alveoli are not perfectly spherical but rather are polyhedral, (2) alveoli change in size and shape, (3) reversible thermodynamics do not strictly apply, and (4) alveoli are not independent units but rather interact with neighboring units. However, the equation offers a reasonable estimate of some of the variables involved.

9.2. PROPERTIES OF DPPC

The physical properties of a surfactant might be expected to be determined mainly by its major constituents, the phospholipids. Since the major phospholipids of surfactant are phosphatidylcholines (PC) containing either two saturated palmitic acids [dipalmitylphosphatidylcholine (DPPC)] or one monounsaturated oleic acid and one palmitic acid, (oleylpalmitylphosphatidylcholine), many studies have been done on these types of lipids. Both the saturated and unsaturated PCs adsorb to the surface of water in a Langmuir trough and form a compressed monolayer film on the surface when the area of the film is properly compressed. This generates a surface tension–surface area isotherm from which one can calculate the surface tension (q) and molecular surface area per lipid molecule. Each molecule of DPPC at a surface tension of 5 dynes/cm has a molecular area of about 3.82 nm². Since its surface tension changes markedly with a small change in surface area, DPPC has a high surface elastance (Fig. 9-3).

Egg PC, which is a mixture of PCs, forms a more expanded surface film and is unable to reduce surface tension below 28 dynes/cm. On this basis, therefore, DPPC is a better candidate as a surfactant than is egg PC.

Why, then, does the natural lung surfactant contain both saturated and unsaturated PCs as well as other phospholipids and proteins. Part of the answer can be see in Fig. 9-4, which shows the spreading properties of DPPC, lung

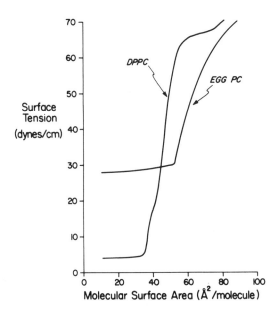

Figure 9-3. Surface tension–surface area isotherms of DPPC and egg PC. The phopholipids were spread from a chloroform–methanol solution on a subphase of 0.10 M NaCl, pH 7.4. The surface was compressed at a rate of 93 cm²/min. (From King, 1978. Reproduced with permission.)

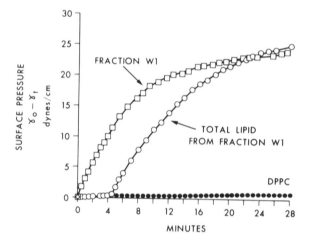

Figure 9-4. Rates of adsorption at 37 °C of lung surface active material (fractoin W1), lipids extracted from surface-active material, and DPPC. Experiments were done in Ringer's solution buffered to pH 7.4. (From King, 1978. Reproduced with permission.)

surfactant (fraction W1), and lipids extracted from lung surfactant. The intact lung surfactant spreads most quickly, indicating a role of the proteins and other lipids in enhancing the spreadability of the surfactant. After a lag period of about 5 min, the surfactant lipid spreads quickly. DPPC alone spreads very poorly. On this basis, DPPC would be a poor surfactant. Thus, a complex interplay between the proteins and lipids of lung surfactant must occur.

9.3. LUNG SURFACTANT

Lung surfactant is complex, containing lipids, proteins and carbohydrate (Table 9-1).

Saturated PCs such as DPPC are effective in lowering surface tension of the alveolar surface and increasing its surface elastance. To do so the surfactant must be rapidly spreadable. This is accomplished by packing DPPC with other unsaturated phospholipids and proteins that allow DPPC to spread rapidly over the alveolar surface. It is likely that during breathing, as the alveoli expand and shrink, the lipid composition of surfactant changes such that unsaturated PCs (and other lipids) are forced out of the surface at small alveolar surface areas. This enriches the film with DPPC and prevents alveolar collapse. When the alveoli expand, the opposite may occur. Thus, at small volumes and small surface area, DPPC is the important surfactant whereas at large volumes and large surface area, the unsaturated PCs may be the effective surfactants.

Table 9-1. Composition of Pulmonary Surfactant Purified from Canine Lavage Fluid[a]

Lipid	Percent of total
Saturated phosphatidylcholines[b]	45
Unsaturated phosphatidylcholines	25
Phosphatidylglycerol	5
Phosphatidylethanolamine	3
Neutral lipids	10
Other phospholipids	2
Proteins	8
Carbohydrate	2

[a]From King, 1978. Reproduced with permission.
[b]Rich in DPPC.

The complex surfactant is able to adjust its composition and accommodate the surface tension and surface elastance during breathing. What, then, is the role of the proteins? It is likely that the proteins are required to help store the surfactant in lamellar bodies of type II cells (pneumocytes), regulate the release of the phospholipid surfactant, and enhance the spreadability of the phospholipids, especially DPPC.

Lung surfactant is synthesized by type II cells. Fig. 9-5 depicts the secretion of surface-active material into the alveolar space and its absorption to alveolar interface. Type II alveolar cells produce inclusion bodies that contain surfactant. The inclusion bodies migrate to the alveolar surface, where they undergo transformation to tubular myelin structures, which provide the phospholipid that is adsorbed as a monolayer at the alveolar–air interface. Type II cells also can take up the secreted lipids.

Lung surfactant consists of several types of phospholipids. The lipid composition varies among individual animals and humans and also varies with age, especially in the developing fetus and immediately after birth (Fig. 9-6). The PC content increases and the sphingomyelin (SPH) content decreases.

The marked increase in the PC/SPH ratio is the most dramatic lipid alteration during development. The types of phospholipid that occur in lung surfactant include sphingomyelins, phosphatidylcholines, phosphatidylethanolamines, and phosphatidylglycerols. These phospholipids represent heterogeneous populations of each type since they contain both saturated and unsaturated fatty acids of different chain lengths. As mentioned above, a major phospholipid is DPPC containing two palmitic acids (a linear fatty acid with 16 carbon atoms and no double bonds). At one time, DPPC alone was believed to be the surfactant.

Alveolar   lumen

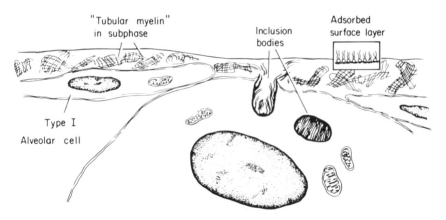

Figure 9-5. Synthesis and secretion of lung surfacant. (From King, 1974. Reproduced with permission.)

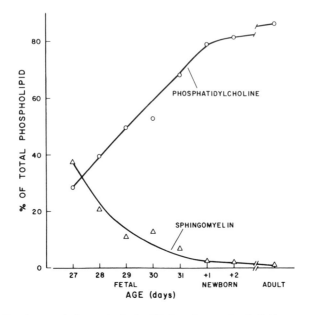

Figure 9-6. Developmental changes in fetal rabbit lung lavage phospholipid composition. (From Rooney, 1978. Reproduced with permission.)

Recent studies have questioned this concept, since a complex interplay between several phospholipids and proteins make up the active surfactant, and DPPC alone has very poor spreadibility at a water–air interface. It is noteworthy, however, that of all animal tissues, lung contains the highest amount of DPPC and has the necessary enzymes required for its synthesis. It is also clear that the amount of DPPC increases during late gestation and rises precipitously immediately after birth. As mentioned above, it is believed that DPPC plays its most important role as a surfactant at small alveolar volumes during expiration and thereby prevents collapse of the lung alveoli.

9.4. SYNTHESIS OF LUNG SURFACTANT PHOSPHOLIPIDS

To understand how DPPC is synthesized by lung cells a knowledge of the functions of phospholipids and their biosynthesis is necessary.

Phospholipids are important constituents of cell membranes in which they make up the lipid bilayer structure. These complex lipids are amphiphiles, since they consist of both a polar and a hydrophobic moiety. They form lipid bilayers spontaneously. Some of these lipids contain polar nonionic head groups, while others contain polar ionic head groups. They are asymmetrically arranged in cell membranes, and each membrane usually has its own characteristic composition and arrangement of lipids.

Phospholipids and glycolipids have a variety of functions in the body. In addition to their major function in cell membranes, they play a role in blood coagulation, in the surface activity of the alveoli of the lungs, in lipoproteins of the blood, and in prevention of water loss in the skin. Their functions in cell membranes include the following: (1) they allow a proper milieu (viscosity) for membrane proteins and enzymes, (2) they form the structure of myelin in nerve cells, (3) they are involved in the conduction of the nerve impulses and secretion of neurotransmitters in the brain, (4) they are important for the structure and function of retinal rod disks in vision, (5) they play a role in cell–cell interactions and communication, and (6) they form a semipermeability barrier on membranes which regulates the flow of ions and nutrients into cells.

The synthesis of phospholipids and glycolipids is complex, involving enzymes both in the cytosol and in the endoplasmic reticulum. The polar water-soluble intermediates are synthesized in the cytosol whereas the the hydrophobic moieties are synthesized and assembled on the endoplasmic reticulum membrane.

Phospholipids represent a variety of phosphorylated lipids that differ in fatty acid composition and in the polar organic molecule attached to the phosphate group. These polar molecules are choline, serine, ethanolamine, inositol, inositolphosphates, and glycerol. Cytidine nucleotide intermediates are required for the transfer of groups such as choline, ethanolamine, and diacylglycerol during the synthesis of phospholipids.

The fatty acids that are esterified to the glycerolphospholipids are asymmetrically positioned. Saturated fatty acids occur primarily on the alpha (C-1) position, whereas unsaturated fatty acids occur primarily on the beta (C-2) position of the glycerol backbone.

The most abundant phospholipids in animal cells are phosphatidylcholine, phosphatidylethanolamine, phosphatidylserine, phosphatidylinositols, and cardiolipin (diphosphatidylglycerol). Cardiolipin occurs primarily on the inner mitochondrial membrane.

The metabolism of phospholipids and glycolipids is complex involving many enzymes and cofactors. An important intermediate in the synthesis of glycerolphospholipids is phosphatidic acid. It is formed by a two-step acylation of either (1) sn-glycerolphosphate or (2) acylation of dihydroxyacetonephosphate (Fig. 9-7). Phosphatidic acid also can be produced by phosphorylation of diacylglycerol, a reaction catalyzed by diacylglycerol kinase (Fig. 9-8).

The acylation reactions show some specificity, since unsaturated fatty acids are esterified on the C-2 position and saturated fatty acids are esterified on the C-1 position of glycerolphosphate. In the intestinal mucosa, 1-monoacylglycerol is phosphorylated to lysophosphatidic acid, which is then acylated to phosphatidic acid (Fig. 9-9).

Since phosphatidic acid is readily converted to other products, its concentration in cells is very low. This made its detection very difficult but was finally resolved by use of high levels of radioactive ^{32}P-labeled inorganic phosphate or ^{32}P-labeled ATP in intact animals or cell homogenates, respectively.

The interaction of diacylglycerol with CDP-choline or CDP-ethanolamine leads to the synthesis of PC and phosphatidylethanolamine (PE) respectively (Fig. 9-10). These are the two most abundant phospholipids in mammalian cells.

(A) Acylation of glycerolphosphate

Glycerolphosphate + acyl-CoA ---> lysophosphatidic acid + CoA

Lysophosphatidic acid + acyl-CoA -----> phosphatidic acid + CoA

(B) Acylation of dihydroxyacetonephosphate (DHAP)

DHAP + acyl-CoA ---------------> acyl-DHAP + CoA

Acyl-DHAP + NADPH + H$^+$ ---------> lysophosphatidic acid + NADP$^+$

Lysophosphatidic acid + acyl-CoA ----> phosphatidic acid + CoA

Figure 9-7. Synthesis of phosphatidic acid.

Diacylglycerol + ATP -------> phosphatidic acid + ADP

Figure 9-8. Phosphorylation of diacylglycerol.

Monoacylglycerol + ATP ------> lysophosphatidic acid + ADP

Lysophosphatidic acid + acyl-CoA ----> phosphatidic acid + CoA

Figure 9-9. Conversion of monoacylglycerol to phosphatidic acid.

Choline + ATP -------------> cholinephosphate + ADP

Cholinephosphate + CTP --------> CDP-choline + PPi

CDP-choline + diacylglycerol ----> phosphatidylcholine + CMP

Ethanolamine + ATP ------> ethanolaminephosphate + ADP

Ethanolaminephosphate + CTP -----> CDP-ethanolamine + PPi

CDP-ethanolamine + diacylglycerol --> phosphatidylethanolamine + CMP

Figure 9-10. Synthesis of phosphatidylcholine and phosphatidylethanolamine.

The rate-limiting step in their synthesis is the reaction of phosphorylcholine and phosphorylethanolamine with CTP to form CDP-choline and CDP-ethanolamine. The pathway for the synthesis of PC that utilizes CDP-choline is referred to as the salvage pathway since it uses (salvages) dietary choline.

PC also can be synthesized by methylation of PE. This reaction occurs primarily in liver and utilizes the methyl donor S-adenosylmethionine. Apparently, two methylating membrane-bound enzymes are required. These enzymes are asymmetrically arranged in the membrane and may be activated when certain hormones, such as epinephrine, bind to their membrane receptors. The activation of these methyl transferase enzymes, which converts PE to PC, is postulated to modify the membrane fluidity in a specific domain where the hormone receptor resides and thus allows for enhancement of signal transduction by the activated receptor. This finding has yet to be confirmed.

The syntheses of phosphatidylserine, phosphatidylinositol, phosphatidylglycerol, and cardiolipin all involve the participation of CDP-diacylglycerol (Fig. 9-11).

It is readily apparent that cytidine nucleotide derivatives play a vital role in the synthesis of phospholipids.

A unique feature of lung surfactant is that it contains a high amount of the saturated DPPC. How this saturated phospholipid is synthesized has been of particular interest, since it is an important constituent of lung surfactant. A deficiency of lung surfactant is a common problem in premature infants, giving

Phosphatidic acid + CTP -------> CDP-diacylglycerol + CMP
(CDP-DAG)

CDP-DAG + serine ---------> phosphatidylserine + CMP

CDP-DAG + inositol -------> phosphatidylinositol + CMP

PI + ATP -------> phosphatidylinositol-4-phosphate + ADP

PIP + ATP -------> phosphatidylinositol-4,5-bisphosphate + ADP

CDP-Diacylglycerol + GP -------> phosphatidylglycerol + Pi + CMP
(PG)

CDP-Diacylglycerol + PG -----> diphosphatidylglycerol + CMP
(cardiolipin)

Figure 9-11. Synthesis of phospholipids utilizing CDP-diacylglycerol as an intermediate.

rise to RDS of the newborn and in some cases leading to death. Synthesis of PC by the usual salvage pathway (Fig. 9-10) produces mixed PCs containing both saturated and unsaturated fatty acids. Therefore the synthesis of disaturated DPPC must involve separate synthetic pathways by Type II cells. Type II cells have a great capacity to take up palmitic acid from the blood.

Synthesis of DPPC has been shown to occur by three different pathways (Fig. 9-12). In pathway 1 (the deacylation-acylation pathway) the action of phospholipase A_2 removes the unsaturated fatty acid on the C-2 position of PC

Pathway 1

phospholipase A_2
PC   --------------------> lyso-PC + fatty acid
mixed C16   C16 sat.   unsat.
sat. + unsat.

Lyso-PC + palmityl-CoA -------> dipalmityl-PC + CoA
C16 sat.

Pathway 2

LLAT
Lyso-PC + Lyso-PC --------> dipalmityl-PC + GPC
C16 sat.   C16 sat.

Pathway 3

GPC + 2palmityl-CoA --------> dipalmityl-PC + 2CoA

Figure 9-12. Synthesis of DPPC.

and forms a saturated palmityl-lyso-PC, which then reacts with a palmityl-acyl-CoA to form DPPC. In pathway 2, (the deacylation-transacylation pathway) two lyos-PCs formed by the action of phospholipase A_2 undergo a dismutase reaction with the enzyme lyso-PC-lyso-PC acyl transferase (LLAT) to form DPPC and glycerolphosphorylcholine (GPC). LLAT was discovered by Erbland and Marinetti (1965a,b). Pathway 3 has recently been postulated by Infante and Huszagh (1987) based on the evidence that lung has a high content of GPC and has a high GPC synthase activity. Pathway 1 appears to predominate in Type II cells.

Increasing amounts of lung tissue and surfactant phospholipids are produced by the fetal lung as gestation progresses. The rise in phospholipid content of lung tissue is due mainly to the rise in PC. The amount of saturated PC also increases. These increases in phospholipid content and composition are more dramatic in lung surfactant and lamellar bodies. There is also an increase in the amount of phosphatidylglycerol but a decrease in the amount of phosphatidylinositol. Indeed, the rise in phosphatidylglycerol in human surfactant can be used as a marker of lung maturity. The rise in these phospholipids results from an increased biosynthesis and secretion by type II cells and may also result from an increase in the number of type II cells. Studies indicate that the activities of phosphorylcholine-CTP transferase (CYT) and lyso-PC acyl transferase increase as gestation progresses. The increase in lyso-PC acyl transferase is much greater than that of LLAT. Some studies indicate that LLAT activity does not increase during gestation. Therefore there is some conflict as to the importance of LLAT in DPPC synthesis.

Thus, the increased synthesis of DPPC depends on the level and activity of the various enzymes involved and the number of type II cells but also depend on the level of palmitic acid. In turn, the amount of each enzyme present is related to its rate of synthesis, which is under hormonal regulation. Moreover, activators and inhibitors may be involved, and enzyme modification by phosphorylation–dephosphorylation must be considered in the regulation of phospholipid synthesis. These factors, plus the fact that the lung contains different cell types, one or more of which might be able to synthesize surfactant, make the study of surfactant a complex problem. It is generally believed that type II cells produce all the surfactant in lung.

9.5. HORMONAL REGULATION OF THE SYNTHESIS OF LUNG SURFACTANT

Some enzymes required for PC synthesis are increased in either amount or activity just before and just after birth. An important question is which enzyme is rate-limiting and most subject to hormonal regulation and increased synthesis. The rate-limiting step in the synthesis of PC by the salvage pathway is catalyzed by CYT. The synthesis of CYT is increased by hormonal manipulation (by glucocorticoids and estrogens), as shown by studies on fetal rabbits. Dexamethasone also increases CYT in fetal lung in organ culture. CYT is stimulated by phosphatidylglycerol in vitro experiments. This fact is particularly interesting since phosphatidylglycerol increases in amount in lung tissue and surfactant just before birth.

Glucocorticoids and estrogen have been reported to increase the activity or synthesis of lysolecithin-lysolecithin acyl transferase, lysolecithin acyl transferase, and phosphatidic acid phosphatase. However, other research has not confirmed these results. Thus, although it is clear that hormones such as a glucocorticoids and estrogen stimulate the synthesis of lung surfactant before or just after birth, it is not clear how this is accomplished. Some studies indicate that glucocorticoids stimulate surfactant lipid synthesis in type II cells indirectly by stimulating fibroblasts to elaborate a diffusible factor that stimulates type II cells.

Thyroid hormones may directly act on type II cells by increasing their responsiveness to the diffusible factor elaborated by fibroblasts. Available evidence suggests that hormones stimulate the synthesis of surfactant DPPC by activation of cholinephosphate cytidyl transferase. Glucocorticoids and thyroid hormones may also have a role in the increased synthesis of β-adrenergic receptors in type II cells that occurs in late fetal life.

Lung surfactant contains not only phospholipids but also proteins that are vital for its properties. Two major protein groups have been identified, a glycoprotein (SP-28-36) with a molecular weight of 26–40 kDa and several hydrophobic proteins (SP-18 and SP-5) with molecular weights of 5 and 18 kDa respectively. How the proteins function and their molecular structure remain to be determined. Phospholipid synthesizing enzymes and hormonal regulation of lung surfactant are covered in a recent review (Post and van Golde, 1988).

The study of lung surfactant and RDS has attracted the attention of many researchers in various fields, in particular biophysicists. However, the prevention and treatment of RDS has been difficult to manage. The use of better ventilation therapy has increased the survival rate of infants afflicted with RDS. Assay of the PC/SPH ratio in amniotic fluid has been helpful in predicting which infants are likely to be affected and how severely. The use of glucocorticoid therapy administered to mothers in premature labor in order to enhance fetal lung maturity and surfactant levels is now commonly used. Despite of these advances, many infants develop RDS, and many who survive suffer complications of bronchopulmonary dysplasia later in life.

Since the surfactant develops late in gestation or immediately after birth, premature babies are most susceptible to RDS. The surface tension of lung extracts obtained from babies with RDS is abnormally high, resulting in alveolar lability, alveolar collapse, and impaired lung function. Alveolar ventilation is so reduced that asphyxia, hypoxemia, and acidemia occur. Vasoconstriction also occurs under these conditions, resulting in a decreased blood flow to poorly ventilated alveoli. This leads to a further decrease in synthesis of surfactant and aggravates an existing serious condition. In severe cases, the air–blood barrier is ruptured and effusion into the alveoli results. Fibrous material (hyaline) may be formed, due in part to blood clotting. These conditions ultimately lead to death.

9.6. THERAPY FOR RDS

The serious nature and prevalence of RDS has led a number of researchers to prepare a synthetic surfactant. Early attempts were unsuccessful and in many cases did more harm than good. Robert Notter and co-workers (Notter and Shapiro, 1981), at the University of Rochester, and other laboratories have prepared artificial multicomponent surfactants by mixing different types of lipids with and without the protein components isolated from lung surfactant. Recently, commercial preparations of artificial surfactant have been developed. These artificial surfactants are hoped to be effective in helping infants with RDS. The proteins may present a problem, since they can be antigenic. The phospholipids used must be protected from peroxidation and must be mixed in the best molecular ratio, which has yet to be determined. Cholesterol addition, if necessary, cannot be at too high a level, since cholesterol can decrease the surface activity of DPPC. The proper physical form of the mixture will vary with temperature and composition and can be a serious problem when the artificial surfactant is sterilized. Finally, the best method for delivery to the infant (tracheal instillation or aerosolization) must be determined.

In summary, lung surfactant is a complex lipoprotein that lowers and adjusts surface tension in the lung as a result of its rapid adsorption and dynamic spreadability on the alveolar surface. Lung surfactant is synthesized in the lamellar bodies of type II cells (pneumocytes). The lamellar body of pneumocytes is an intracellular organelle composed of lamellae of lipids and proteins. These bodies are secretory organelles that release surfactant by exocytosis, a process stimulated by such hormones as epinephrine, acetylcholine, thyroxine, and steroid hormones. The proteins may stabilize the lamellae and control the rate of release of the surfactant into the alveoli. Steroid hormones (glucocorticoids) and estrogens stimulate the production of lung surfactant just before birth and after birth. The major functions of surfactant in vivo are to decrease the work of breathing, increase lung compliance, possibly decrease the driving force for pulmonary edema, and stabilize alveoli and prevent atelectasis. A deficiency of lung surfactant in the newborn results in RSD, a potentially fatal and debilitating condition afflicting about 25,000 premature babies each year.

REFERENCES

Avery, M. G., and Mead, J., 1959, Surface properties in relation to atelectasis and hyaline membrane disease, *Am. J. Dis. Child.*, 97:517.

Clements, J. A., 1970, Pulmonary surfactant, *Am. Rev. Respir. Dis.*, 101:984.

Clements, J. A., 1974, Biochemical aspects of pulmonary function, *Fed. Proc.*, 33:2231.

Erbland, J., and Marinetti, G. V., 1965a, The enzymatic acylation and hydrolysis of lysolecithin, *Biochim. Biophys. Acta.*, 106:128

Erbland, J., and Marinetti, G. V., 1965b, The metabolism of lysolecithin in rat liver particulate systems, *Biochim. Biophys. Acta*, 106:139.

Farrell, P. M., and Avery, M. E., 1975, Hyaline membrane disease, *Am. Rev. Respir. Dis.*, 111:657.

Hawthorne, J. N., and Ansell, G. B. (eds.), 1982, *Phospholipids*, Elsevier, Amsterdam, The Netherlands.

Infante, J. P., and Huszagh, V. A., 1987, Is there a new biosynthetic pathway for lung surfactant phosphatidylcholine, *Trends Biochem. Sci.*, 12:131.

King, R. J., 1974, The surfactant system of the lung, *Fed. Proc.*, 33:2254.

King, R. J., 1979, Pulmonary surface active material: basic concepts, in *The Surfactant System and the Neonatal Lung*, Mead Johnson Symposium on Perinatal and Developmental Medicine, published by Mead Johnson & Co., Evansville, Indiana, No. 14, p. 3.

King, R. J., Carmichael, M. C., and Horowitz, C. M., 1983, Reassembly of lipid-protein complexes of pulmonary surfactant: proposed mechanism of interaction, *J. Biol. Chem.*, 258:10672.

Morgan, T. E., 1971, Pulmonary surfactant, *N. Engl. J. Med.*, 284:1185.

Notter, R. H., and Shapiro, D. L., 1981, Lung surfactant in an era of replacement therapy, *Pediatrics*, 68:781.

Pelech, S. L., and Vance, D. E., 1984, Regulation of phosphatidylcholine biosynthesis, *Biochim. Biophys. Acta*, 779:217.

Post, M., and van Golde, L. M. G., 1988, Metabolic and developmental aspects of the pulmonary surfactant system. *Biochim. Biophys. Acta*, 947:249.

Rooney, S. A., 1975, Development of pulmonary surfactant system during late fetal and early postnatal life, *Am. Rev. Respir. Dis.*, 111:657.

Rooney, S. A.,1979, Pulmonary surface active material: basic concepts, in *The Surfactant System and the Neonatal Lung*, Mead Johnson Symposium on Perinatal and Developmental Medicine, published by Mead Johnson & Co., Evansville, Indiana, No. 14, p.17.

Smith, B. T., 1979, The surfactant system and the neonatal lung, in *Biochemistry and Metabolism of Pulmonary Surface-active Material*, Mead, Johnson Symposium on Perinatal and Developmental Medicine, published by Mead Johnson & Co., Evansville, Indiana, No. 14, p. 12.

Smith, E. L., Hill, R. L., Lehman, I. R., Lefkowitz, R. J., Handle, P., and White, A., 1983a, *Principles of Biochemistry: Mammalian Biochemistry*, 7th ed., McGraw-Hill Book Co., New York.

Smith, E. L., Hill, R. L., Lehman, I. R., Lefkowitz, R. J., Handler, P., and White, A., 1983b, *Principles of Biochemistry: General Aspects*, 7th ed., McGraw-Hill Book Co., New York.

Staub, N. C., 1970, The pathophysiology of pulmonary edema, *Hum. Pathol.*, 1:419.

Tierney, D. F., 1974, Intermediary metabolism of the lung, *Fed. Proc.*, 33:2232.

Von Neergaard, K., 1929, Neue auffassungen uber einen grundbegriff der atemmechanik, *Z. Ges. Exptl. Med.*, 66:373.

Chapter 10
SPHINGOLIPIDOSES
Gangliosidoses, Tay-Sachs Disease, and Sandhoff's Disease

10.1. EARLY HISTORY

Sphingolipids occur in high amounts in brain and nerve cells, where they play important roles in membrane structure and function. Disorders in the degradation of sphingolipids give rise to a variety of genetic diseases called the sphingolipidoses. The gangliosidoses represent a subclass of the sphingolipidoses. In 1881, Warren Tay, a British ophthalmologist, first described a genetic abnormality of ganglioside metabolism. The case he described is now known as Tay-Sachs disease or GM_2 (ganglioside M_2) gangliosidosis. Tay-Sachs disease (TSD) is the most common ganglioside disease known. The heterozygote frequency among Ashkenazi Jews is 1 in 27 persons. The disease can be determined by assay for the defective enzyme in blood. Because of the high frequency of the disease in high-risk populations, mass screening programs have been carried out in many cities in 13 countries with more than 312,000 persons tested since 1970. Over 250 at-risk carrier couples with no history of TSD in their families have been identified. Prenatal diagnosis of affected fetuses by amniocentesis and enzyme assay is the only preventive solution available for this fatal disease.

In infantile TSD, onset is within the first year of life and death occurs in early childhood, after progressive motor and mental retardation. In the adult disease, onset usually occurs in the second or third decade of life and is characterized by lower motor neuron, pyramidal tract, and cerebellar deterioration.

10.2. GENETIC DEFECTS

The molecular genetics of the enzymes hexosaminidase A (hexA) and hexosaminidase B (hexB) which are involved in TSD have been elucidated. These

205

enzymes hydrolyze an N-acetylhexosamine (galactosamine) from the ganglioside during its normal degradation. If the enzyme is defective, the GM_2 ganglioside accumulates in affected cells and leads to the symptoms of the disease. Both enzymes have molecular weights of 100,000. Only hexA cleaves ganglioside GM_2, whereas hexB can cleave other substrates such as globoside, containing N-acetylhexosamine.

HexA consists of two subunits, an α-chain encoded by a gene locus on chromosome 15 and a β-chain encoded by a gene locus on chromosome 5. HexB is a tetrameric protein containing four β subunits each having a molecular weight of 25,000 and is designated $\beta_2\beta_2$. An activator protein is required only for hexA to function. If this protein is defective or deficient, it leads to AB(+) GM_2 gangliosidosis. A large number of hexA and hexB variants have now been described, and a nomenclature based on a three-gene locus model has been proposed to classify these variants. In this model mutations at the α locus on chromosome 15 give rise to hexA deficiency, causing TSD (juvenile and adult GM_2 gangliosidosis). Mutations at the β locus on chromosome 5 lead to hexA and hexB deficiency, causing infantile, juvenile, or adult Sandhoff's disease. Mutations at the activator locus lead to AB(+) GM_2 gangliosidosis (Fig. 10-1). Recent work by Navon and Proia (1989) on Ashkenazi adult GM_2 gangliosidosis patients from five families has revealed a point mutation in the α-chain gene coding for β-hexA that results in the substitution of glycine-269 for serine. This amino acid substitution drastically depresses the catalytic activity of the enzyme.

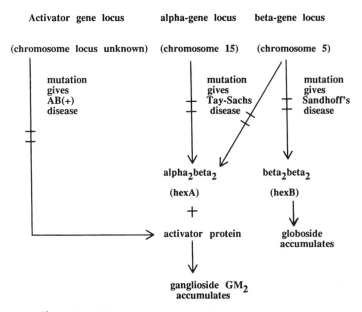

Figure 10-1. Mutations leading to Sandhoff's disease and TSD.

10.3. SPHINGOLIPIDS

An understanding of the sphingolipidoses requires a knowledge of the biochemistry of the sphingolipids. Sphingolipids include phosphorylated derivatives of ceramides (sphingomyelin), and nonphosphorylated derivatives of ceramides containing different types and number of carbohydrate residues. Ceramides, which are fatty acyl amides of sphingosine, include, cerebrosides, sulfatides, glycosylceramides, and gangliosides. The fatty acids in the sphingolipids are linked via an amide bond to the amino group of sphingosine. Sphingolipids occur in high amounts in brain and nerve cells. Gangliosides are of two types, sialo and nonsialo. Sialogangliosides contain hexoses, hexosamines, and sialic acid (neuraminic acid) linked to the terminal hydroxyl group of sphingosine. Ganglioside GM_3 lacks the terminal N-acetylgalactosamine, whereas ganglioside GM_1 contains one more galactose unit linked β-1,3 to the terminal N-acetylgalactosamine of GM_2. Gangliosides containing two neuraminic acids are designated GD, and those containing three neuraminic acid residues are designated GT. Gangliosides lacking neuraminic acid are called asialogangliosides, designated as GA. The gangliosides are found in high concentrations in brain and nerves and their concentrations vary with age and location in the brain.

10.3.1. Biosynthesis of the Sphingolipids

The synthesis of sphingolipids is shown in Fig. 10-2. Ceramide and CTP play an important role in the synthesis of sphingolipids. Many of the enzymes involved are localized in the endoplasmic reticulum. The sphingolipids are important for the synthesis of neuronal cell membranes and therefore are vital constituents of brain and nerve.

There is considerable variation in the structures of the ceramide and hydrocarbon chains of the glycosphingolipids. About 130 varieties of these lipids are now known. About 40 are assigned to the ganglio class, 10 are of the globo class, and 60 are of the lacto class; this classification is based on the sequence of sugars found in the core of the carbohydrate chains and the nature of the chemical linkages between the sugars.

The first glycosphingolipid was discovered in brain tissue by Thudichum in 1874 and was named cerebroside. Gangliosides were discovered by Klenk much later, in 1936. Klenk found gangliosides only in brain tissue, but other researchers have since found them in much smaller amounts, in all other animal tissues. Gangliosides are characterized by the presence of one or more acidic sugars called sialic acid (neuraminic acid). GM_1 is the specific receptor for the tetanus toxin as well as the receptor for the cholera toxin. Gangliosides also interact with other biologically important substances such as botulinum toxin, interferon, interleukin, serotonin, hormones, and Sendai virus.

Glycosphingolipids are found in all animal cells and in some plant cells. Blood group antigens and many other antigens that are modified in the develop-

(A) Synthesis of sphingomyelin

Sphingosine + fatty acyl-CoA -------> ceramide + CoA

Ceramide + CDP-choline -------------> sphingomyelin + CMP

(B) Synthesis of gangliosides

Ceramide-glc + UDP-gal ------> ceramide-glc-gal + UDP

Ceramide-glc-gal + CMP-NANA -------> ceramide-glc-gal + CMP
                                                      |
                                                    NANA
                                            (GM$_3$ ganglioside)

Ceramide-glc-gal + UDP-galNAc ----> ceramide-glc-gal-galNAc + UDP
            | |
          NANA   NANA
                                            (GM$_2$ ganglioside)

Ceramide-glc-gal-galNAc + UDP-gal --> ceramide-glc-gal-galNAc-gal + UDP
            | |
          NANA   NANA
                                    (GM$_1$ ganglioside)

(C) Synthesis of cerebrosides and sulfatides

Ceramide   +   UDP-galactose -------> galactocerebroside + UDP

Ceramide + UDP-glucose ----------> glucocerebroside + UDP

Sphingosine + UDP-galactose -----> sphingosine-galactose + UDP

Sphingosine-galactose + acyl-CoA ----> galactocerebroside + CoA

Sphingosine + UDP-glucose -------> sphingosine-glucose + UDP

Sphingosine-glu + acyl-CoA -------> glucocerebroside + CoA

Galactocerebroside + PAPS --------> cerebrosidesulfate + PAP
                                              (sulfatide)

Figure 10-2. Synthesis of sphongolipids. CTP, cytidine triphosphate; CMP, cytidine monophosphate; UDP, uridine diphosphate; NANA, N-acetylneuraminic acid; NAc, N-acetyl; glc, glucose, gal, galactose; galNAc, N-acetylgalactosamine; PAPs, phosphoadenosinephosphosulfate.

ment of malignant tumors are glycosphingolipids. Glycosphingolipids are localized on the outer plasma membrane of cells and are believed to mediate the interactions of cells with their environments. They serve as distinguishing markers for cells from various organs in an animal and mediate cell–cell recognition and communication. They regulate the growth of cells by sensing the density of similar cells in their vicinity and can inhibit the response of the cell to growth factors.

Glycosphingolipid antigens such as the blood group substances are secondary gene products. Their structures are not encoded in DNA, as are the sequences of amino acids in proteins. Instead, a glycosphingolipid is synthesized in a series of reactions catalyzed by multiple enzymes called glycosyl transferases. The sugar sequence is determined by the glycosyl transferases in the membrane.

10.3.2. Enzymatic Degradation of the Gangliosides

The gangliosides are broken down stepwise by specific hydrolases beginning at the nonreducing end of the oligosaccharide chain (Fig. 10-3). This scheme is a simplified version of a more complex scheme involving the higher gangliosides GD and GT, which contain two and three neuraminic acid residues, respectively. Each gangliosidosis appears to be genetically distinct and is transmitted as an autosomal recessive trait. Since the enzymes involved in the degradation of the gangliosides occur in lysosomes, these diseases are known as lysosomal enzyme deficiency diseases. Moreover, since the degradation products accumulate in lysosomes that become engorged, these conditions are characterized as lysosomal storage diseases.

10.3.3. Distribution of Gangliosides in the Brain

GM_1 occurs in highest amounts in the brains of both newborns and adults. In contrast, GM_3 is a minor constituent in gray and white matter of the brain of the adult and newborn.

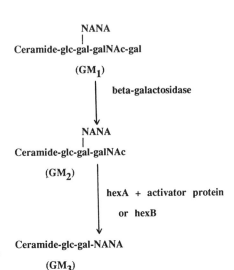

Figure 10-3. Degradation of gangliosides. Abbreviations are as for Fig. 10-2.

10.3.4. Types and Clinical Features of GM$_2$ Gangliosidoses

The various types of GM$_2$ gangliosidoses, their years of discovery, and number of reported patients shows that TSD was the first to be discovered in 1881 and is the most prevalent, with thousands of cases now known. Sandhoff's disease, juvenile GM$_2$ gangliosidosis, and the AB(+) variant were discovered in 1968, and only 5–100 patients have been reported. Adult GM$_2$ gangliosidosis was discovered in 1973, and approximately 10 cases are known.

The major clinical features of the GM$_2$ gangliosidoses include mental motor retardation, neuronal lipidosis, macrocephaly, seizures, and blindness. In TSD, Sandhoff's disease, and juvenile GM$_2$ gangliosidosis, onset of symptoms occurs early in life, from 3 months to 2–6 years. TSD disease and Sandhoff's disease are the most lethal. Both are associated with a cherry-red spot in the retina (a result of white lipid accumulation in areas of the macula around the red spot), seizures, mental retardation, macrocephaly, hypotonia, neuronal lipidosis, and doll-like facial expression. In these diseases, motor weakness usually begins between 3–6 months of age. The startle reaction, an extension response to sudden, sharp sounds, is a characteristic early symptom. Mental and motor deterioration develop rapidly after 1 year. After 18 months deafness, blindness, convulsions, and spasticity become worse, and a stage of decerebrate rigidity is reached.

Brain levels of GM$_2$ are increased 100–300 times normal in TSD and Sandhoff's disease. The visceral organs do not accumulate GM$_2$, making this form of gangliosidosis a neuronal disease. Changes in other gangliosides and mucopolysaccharides may also be present if these substances contain hexosamines that are susceptible to hexA or hexB hydrolysis. HexA has a molecular weight of 108,000, and hexB has a molecular weight of 100,000. Both have a pH optimum of 4.4. The proposed subunit structure of HexA is $\alpha\beta_2$ and that of Hex B is $\beta\beta_2$. The molecular weight of the α chain is 58,000 and that of the β chain is 25,000.

HexA and hexB hydrolyze both N-acetylglucosamine and N-acetylgalactosamine on gangliosides or mucopolysaccharides that have these groups linked in the β configuration. HexA is more negatively charged and more heat labile than hexB. Both enzymes are found in all normal human tissues. In TSD, hexA activity is nearly absent, but it is of interest that hexB activity may be increased about 10-fold over normal levels. Patients with juvenile GM$_2$ have a partial deficiency of hexA.

The assay for hexA may be misleading depending on the substrate used. With synthetic substrates such as the p-nitrophenyl-4-methylumbelliferyl derivative of either D-N-acetylglucosamine or D-N-acetylgalactosamine, some patients show either a total deficiency of hexA and some show nearly normal activity. However, when the natural GM$_2$ ganglioside is used as substrate, all patients show a nearly total absence of enzyme activity. Part of this difference may be attributable to the physical state of the GM$_2$ substrate and the level of activator protein present. The activator protein appears to be necessary for the dispersion of the ganglioside so that it can react with hexA. The activator protein is not required to disperse the water-soluble synthetic substrates. Persons heterozygous for TSD have intermediate levels of hexA in serum, leukocytes, and cultured fibroblasts.

Sandhoff's disease is characterized by a deficiency in both hexA and hexB. The gene defect is located on chromosome 5. The deficiency occurs in all tissues of the body, including amniotic cells from affected fetuses. The diagnosis is made by assay of hexA and B in the serum or fibroblasts of homozygote patients. Low levels of both enzymes are characteristic of this disease, whereas patients with TSD are deficient only in hexA.

The molecular genetics of GM_2 gangliosidosis suggest a three-gene locus coding for two separate polypeptide chains, α and β, of hexA and for the activator protein. The α chain is mutated in hexA mutants (TSD, juvenile GM_2 gangliosidosis, and adult GM_2 gangliosidosis). The β chain is mutated in hexB mutants (Sandhoff's disease and juvenile Sandhoff's disease). The activator protein is mutated in the AB(+) mutant. This model explains how the mutation in Sandhoff's disease leads to a loss of both hexA and hexB activity and how restoration of hexA activity may occur in fusion of TSD fibroblasts with Sandhoff's disease fibroblasts. The normal α subunit may be derived from fibroblasts of Sandhoff's disease patients, with the normal β subunit being derived from fibroblasts from TSD patients.

Recent work has indicated that the activator gene locus may involve more than one polypeptide, since two separate protein activator proteins have been isolated from human liver. The hydrolysis of GM_1 was stimulated only by a GM_1-specific activator protein that had little activity for the hydrolysis of GM_2. Antiserum raised against the GM_1 activator protein did not cross-react with the GM_2 activator protein and vice versa. In GM_1 gangliosidosis (also called generalized gangiosidosis), the defect involves a β-galactosidase. Therefore, at least two types of hydrolases (β-galactosidase and hexosaminidase) require activator proteins to degrade these ganglioside substrates.

Other types of genetic defects are known that involve the more general class of lipids called sphingolipids (sphingomyelin, cerebrosides, sulfatides, and gangliosides). The general term sphingolipidosis is used to denote defects involving sphingolipids. Gangliosidosis is one such class of sphingolipidosis. The sphingolipidoses, including the defective enzymes and some clinical symptoms are shown in Fig. 10-4, which is a simplified presentation of an increasing number of variants in each disease. Genetic defects are proving to be complex, since several different types of mutation can occur, leading to different alterations in the structure of the polypeptide chain or to a complete loss of the protein.

10.4. OVERVIEW OF THE SPHINGOLIPIDOSES

Some of the sphingolipidoses involve mainly the central nervous system and are localized in either gray matter or white matter, whereas others involve only the visceral tissues or both the visceral tissues and the central nervous system. TSD and Sandhoff's disease involve primarily the central nervous system, which accounts for their clinical symptoms.

The sphingolipidoses represent a class of lysosomal enzyme deficiencies leading to the accumulation of sphingolipids in the lysosomes. Other lysosomal storage diseases afflicting humans are the glycogen storage diseases and the mucopolysaccharidoses. In mixed human populations, lysosomal storage diseases occur at a rate of about 1–5 per 10,000 births. At least 30 diseases of this type are now known, many involving lipid degrading hydrolases and glycosidases.

Each sphingolipid is constantly being synthesized and degraded. When the degradative enzymes are mutated, they cannot function at all or function improperly. This leads to an accumulation of the sphingolipid that they should degrade. The sphingolipid accumulates in the cell and causes cell damage or cell death and gives rise to the specific genetic disorder. The type of first sugar of the sphingolipid defines the localization within the central nervous system. Gray matter contains the ceramide glucoside series, and white matter contains the ceramide galactoside series. Visceral storage of the particular sphingolipid relates to glucoside-type storage whereas central nervous system storage relates to hexosamine storage, the hexosamine being localized early in the oligosaccharide chain.

Sphingolipid degrading enzymes are lysosomal membrane-bound glycoproteins with an acid optimum pH. Many have different subunits and require cofactors for activity.

All sphingolipidoses except Fabry's disease, which is X-linked, show autosomal recessive inheritance. Phenotypically different varieties of many of the sphingolipidoses reflect different enzymatic defects due to different genotypes, depending on the type of mutation. Most of these diseases are fatal in the first few years of life, although some, such as Gaucher's disease and Fabry's disease, persist into adulthood. In these two diseases, the central nervous system is spared.

The sphingolipidoses are rare diseases. Three of the more common ones occur predominantly in Ashkenazi Jews. It is of interest that in contrast to the several genetic diseases involving sphingolipids, no diseases have been reported that involve the glycerolphospholipids other than the deficiency of lung surfactant DPPC, which occurs in RSD of the newborn.

Many lysosomal enzymes are synthesized as large precursor molecules in the rough ER and initially contain a hydrophobic signal peptide of about 20 amino acids. The signal peptide allows the protein to be recognized and incorporated into the membrane of the ER. During its passage from the ER to the Golgi, the signal peptide is removed, carbohydrate units rich in mannose are added, and one or more of these mannose residues are phosphorylated. Phosphorylation is required for binding of the mannose-6-phosphate residue to a specific receptor in the Golgi membrane. The precursor protein is then internalized and further

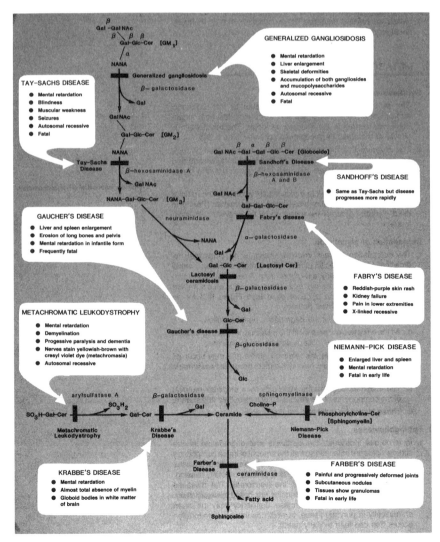

Figure 10-4. Sphingolipidoses: enzyme defects and clinical manifestations. (From Champe and Harvey, 1987. Reproduced with permission.)

processed and finally ends up in the lysosome as the mature active enzyme.

Mutations in genes coding for lysosomal enzymes may cause the following defects: (1) decreased rate of synthesis of the enzyme, with little or no enzyme produced, (2) synthesis of a catalytically inactive mutated enzyme, (3) failure of the mutated enzyme to be transported into the lysosome due in part to the inability to form mannose-6-phosphate, (4) increased rate of degradation of the enzyme, and (5) decreased rate of synthesis of an activator protein.

10.5. THERAPEUTIC APPROACHES FOR THE SPHINGOLIPIDOSES

Therapeutic approaches to genetic diseases must be tailored to the specific disease and depend on the tissue or organ involved and the age of onset. It is obvious that diseases involving the central nervous system will be more difficult to treat than diseases involving organs such as liver and kidney for which organ transplants are possible.

One therapeutic approach is to provide the missing or defective enzyme. To date this approach has not been successful in patients with TSD or Sandhoff's disease and has had temporary modest success in patients with Fabry's disease, since in the latter only the visceral organs are involved. The major problems involved in enzyme replacement therapy are: (1) the isolation of sufficiently pure enzyme from human sources, (2) the ability to direct the enzyme to the specific organ, (3) the lifetime of the administered enzyme, (4) the failure of enzymes to penetrate the blood–brain barrier, (5) production by the patient of antibodies that destroy the administered enzyme, and (6) the high cost of enzyme replacement relative to the small number of affected persons.

Recent advances in tissue transplantation may offer help for certain sphingolipidoses, provided the operation is done early enough and results in production of enough of the normal enzyme in an effective manner.

REFERENCES

Champe, P. C., and Harvey, R. A., 1987, *Lippincott's Illustrated Reviews: Biochemistry*, J. B. Lippincott Co., Philadelphia.

Devlin, T. M. (ed.), 1982, *Textbook of Biochemistry with Clinical Correlations*, John Wiley & Sons, New York.

Li, S. C., Nakamura, T., Ogamo, A., and Li, Y. T., 1979, Evidence for the presence of two separate protein activators for the enzyme hydrolysis of GM1 and GM2 gangliosides, *J. Biol. Chem.* 254:10592.

McGilvery, R. W., and Goldstein, G. W., 1983, *Biochemistry, A Functional Approach*, W. B. Saunders Co., New York.

Navon, R., and Proia, R. L., 1989, The mutations in Ashkenazi jews with adult GM$_2$ gangliosidsosis, the adult form of Tay-Sachs disease. *Science*, 243:1471.

O'Brien, J. S., 1983, The gangliosidoses, in *The Metabolic Basis of Inherited Disease*, (J. B. Stanbury, J. B. Wyngaarden, D. S. Fredrickson, J. L. Goldstein and M. S. Brown, eds., 5th ed., p.945, McGraw-Hill Book Co., New York.

O'Brien, J. S., Okada, S., Fillerup, D. L., Veath, M. L., Adornato, B., Brenner, P. H., and Leroy, J. G., 1971, Tay-Sachs disease: prenatal diagnosis, *Science*, 172:61.

Smith, E. L., Hill, R. L., Lehman, I. R., Lefkowitz, R. J., Handler, P., and White, A., 1983a, *Principles of Biochemistry: Mammalian Biochemistry*, 7th ed., McGraw-Hill Book Co., New York.

Smith, E. L., Hill, R. L., Lehman, I. R., Lefkowitz, R. J., Handler, P., and White, A., 1983b, *Principles of Biochemistry: General Aspects*, 7th ed., McGraw-Hill Book Co., New York.

Stanbury, J. B., Wyngarden, J. B., Fredrickson, D. S., Goldstein J. L., and Brown, M. S., (eds.), 1983, *The Metabolic Basis of Inherited Disease*, pp. 820-972, McGraw-Hill Book Co., New York.

von Figura, K., and Hasilik, A., 1984, Genesis of lysosomal enzyme deficiencies, *Trends Biochem. Sci.*, 9:29.

West, H. H., 1977, The sphingolipidoses, *Postgrad. Med. J.* 61:90.

INDEX

Date Due